Mechanics and Adaptronics

Series Editors

Markus Böl, Institute of Mechanics and Adaptronics, Technische Universität Braunschweig, Braunschweig, Germany

Michael Sinapius, Institute of Mechanics and Adaptronics, Technische Universität Braunschweig, Braunschweig, Germany

Ziel der Buchreihe ist die Publikation neuer Ansätze in der Mechanik und Adaptronik. Diese Ansätze umfassen zum einen experimentelle Themen als auch solche, die numerischen Charakters sind und erstrecken sich über diverse Größenskalen. Experimentelle, zukunfstweisende Themen beinhalten Untersuchungen auf Mikro- bis hin zur Makroebene und berücksichtigen unterschiedlichste Felder, von der Mechanik über Temperatur bis hin zu unterschiedlichen chemischen Feldern. Auch auf Seiten der numerischen Simulation werden auf Grundlage der experimentellen Untersuchungen neue multiskalige Mehrfeldmodelle entwickelt, kalibriert und validiert. Eine Besonderheit stellt hier die Thematik der Adaptronik dar, die als Ziel die Integration neuer Funktionen in tragende Bauteile durch die Kombination konventioneller Konstruktionswerkstoffe mit aktiven Werkstoffsystemen verfolgt. So werden Funktionswerkstoffe betrachtet, die Energiewandler sind und die die klassischen lasttragenden und formgebenden Funktionen um sensorische und aktorische Funktionen erweitern. Aktorische und sensorische Eigenschaften ermöglichen in Kombination mit adaptiven, schnellen Reglern eine optimale Selbstanpassung an die jeweilige Einsatzumgebung. Durch die Zusammenführung von numerischem und experimentellem Know-how auf mechanischer als auch adaptronischer Seite ist es möglich, Erkenntnisse aus der Natur im Sinne des bionischen Gedankens in die Technik zu übertragen und so neue wissenschaftliche Richtungen aufzuschlagen.

Julian Steinmetz

Functional Compliance of Multifunctional Bondlines with Disbond Arrest and Health Monitoring Features for Safe and Efficient Composite Joints

Springer

Julian Steinmetz
Institut für Mechanik und Adaptronik
Technische Universität Braunschweig
Braunschweig, Germany

ISSN 2731-6211 ISSN 2731-622X (electronic)
Mechanics and Adaptronics
ISBN 978-3-031-72421-3 ISBN 978-3-031-72422-0 (eBook)
https://doi.org/10.1007/978-3-031-72422-0

This Springer imprint is published by the registered company Springer Nature Switzerland AG
The registered company address is: Gewerbestrasse 11, 6330 Cham, Switzerland

Functional Compliance of Multifunctional Bondlines with Disbond Arrest and Health Monitoring Features for Safe and Efficient Composite Joints

Von der Fakultät für Maschinenbau
der Technischen Universität Carolo-Wilhelmina zu Braunschweig

zur Erlangung der Würde

eines Doktor-Ingenieurs (Dr.-Ing.)

genehmigte Dissertation

von: Julian Bengt Sebastian Steinmetz
geboren in: Braunschweig

eingereicht am: 11.01.2024
mündliche Prüfung am: 19.07.2024

Vorsitz: Prof. Dr.-Ing. Ulrich Römer
Gutachter: Prof. Dr.-Ing. Michael Sinapius
Prof. C.A. Clemens Dransfeld

2025

To Misato

Preface

This thesis is the result of my work as a research associate at the Institute of Mechanics and Adaptronics (formerly Institute of Adaptronics and Function Integration) of the Technische Universität Braunschweig. It has been conducted in close collaboration with the Institute of Microtechnology of the Technische Universität Braunschweig.

This work was supported by the Deutsche Forschungsgemeinschaft (DFG), grant number 401136681. Hence, before coming to personal acknowledgments, I would like to thank the DFG and the Federal Republic of Germany for enabling our research group and numerous others to pursue research through public funds.

Braunschweig, Germany | Julian Steinmetz
July 2024

Acknowledgements

The greatest joy in my life has been and always will be some outstanding people around me. These are people who, above everything else, have a great heart. I am privileged to say that I have already met many of those people, and it is thanks to them that I could accomplish this thesis. In the following, I would like to express my gratitude to all of you.

First and foremost, I thank you, Prof. Dr.-Ing. Michael Sinapius. You have not only provided me with excellent academic guidance and working conditions but your visions and kindness have also created a very successful team with a unique spirit. From all the lessons you have taught me, the most important one is that excellent leadership means putting the people at the center of your attention. Thank you for always doing that.

I thank Prof. Dr.-Ing. Ulrich Römer for being the head of my examination commission and Prof. C. A. Dransfeld for being the co-supervisor of this thesis. Professor C. A. Dransfeld, thank you for your genuine interest in the topic and for making the discussions pleasant simply by always expressing your joy in research.

In addition, I thank you, Prof. Dr.-Ing. Christian Hühne, for connecting me with the structural bonding group at the German Aerospace Centre, for enabling my research by organizing the funding, and most of all for always being there for the team and me when it matters. Thanks to your commitment, I had the luxury of two academic advisors during my time as a doctoral student, which I appreciate a lot.

I thank the entire team of the Institute of Mechanics and Adaptronics (formerly Institute of Adaptronics and Function Integration). Thank you for all the fruitful discussions, proofreading my work, and making this institute a place where it is fun to work. I am pleased to say that many of you are not just colleagues but good friends. Particularly, I want to thank you, Riem Kilian, Bernd-Christian Hölscher, Patrick Scholle, Maximilian Neitmann, Christian Pommer, and Alexander Kyriazis. Thank you for your academic and moral support, and most of all, for becoming some of my closest friends.

Further, I thank my colleagues Chresten von der Heide from the Institute of Microtechnology, and Dirk Holzhüter, Thomas Löbel, Jens Kosmann, Martin Schollerer, Oliver Völkerink, Robert Prussak and Konstantin Schmidt from the

German Aerospace Centre. I thank you all for being part of the work on multi-functional bondlines, and for your help in conducting this exciting research. Special thanks go to Oliver Völkerink and Chresten von der Heide. Oliver, I thank you for particularly supporting me to finish this thesis, when time was about to run out. Chresten, I thank you for your passion for this project and the numerous micro sensors, whose stepwise improvements led to the successful implementation of the multifunctional bondline in the end. Most of all, I thank you for making the countless hours we shared in the lab not only worthwhile but also fun. I am happy to count you among the great friends I made during my time as a doctoral student.

I also thank all the students who contributed to this work. In particular, I thank you, Nils Vasic, for your contributions regarding the phase transformation of the PVDF material. In this context, I would also like to thank my colleagues Marion Görke and Maik Kahn for your support in these experiments.

I thank my other friends, who coped with me vanishing for some time. In particular, I thank you, Sabine Baumbach and Max Gutierrez, for proofreading my manuscript.

Finally, and most importantly, I thank my family. I am blessed to be part of a family that always supports each other. I thank my parents, Angela and Jörg Steinmetz, for their unconditional love and support throughout my entire life. Lastly, I thank my beloved wife Misato. Misato, I thank you for always having my back. I thank you for bringing Yuki and Aki, the loves of our life, into this world, and taking such good care of all of us.

Braunschweig, Germany Julian Steinmetz
July 2024

Contents

Nomenclature

Abbreviations

CFRP	Carbon Fibre Reinforced Plastic
CLS	Cracked Lap Shear
DAF	Disbond Arrest Feature
DCB	Double Cantilever Beam
DIC	Digital Image Correlation
DSLR	Digital Single-Lens Reflex
EDX	Energy-Dispersive X-ray Spectroscopy
ENF	End Notched Flexure
FBG	Fibre Bragg Grating
FE	Finite Element
FRP	Fibre Reinforced Plastic
ILSS	Interlaminar Shear Strength
MDAF	Multifunctional Disbond Arrest Feature
NDI	Non-destructive Inspection
PEI	Polyetherimide
PTFE	Poly(tetrafluoroethylene)
PVDF	Poly(vinylidene fluoride)
RAPID	Reconstruction Algorithm for Probabilistic Inspection of Damage
SHM	Structural Health Monitoring
SLS	Single Lap Shear
SMT	Surface Mount Technology
VDI	Verein Deutscher Ingenieure e.V.
VSG	Virtual Strain Gauge
VUV	Vacuum Ultraviolet
WSLS	Wide Single Lap Shear
ZIF	Zero Insertion Force

Latin Symbols

A	Absorbance
A_x	Absorbance regarding wavelength x
A_{EA}	Absorbance at any wavelength of an electroactive phase
C_T	Temperature difference
C_s	Coeffcient of strain
E_1	Young's modulus of adherend no. 1
E_2	Young's modulus of adherend no. 2
F_{EA}	Fraction of the electroactive phase
G_A	Shear modulus of the adhesive
H_m	Melt enthalpy
H_{100}	Melt enthalpy of a 100 % crystalline film
K_x	Specific wavenumber regarding wavelength x
L_{min}	Load transfer length
$L_{CLS_{min}}$	Load transfer length of the CLS specimen
P_{amp}	Load amplitude
P_{max}	Maximal load
P_{mean}	Mean load
P_{min}	Minimal load
R	Load ratio
t_1	Thickness of adherend no. 1
t_2	Thickness of adherend no. 2
t_A	Thickness of the adherend
T	Temperature
X_c	Total crystallinity

Greek Symbols

β	Estimate for calculating the load transfer length
β_{CLS}	Estimate for the load transfer length of the CLS specimen
ε	Strain
ε_{max}	Maximum strain
ε_p	Strain of the FBG sensors
ε_x	Strain in the x-direction
ε_y	Strain in the y-direction
λ_b	Measured Bragg wavelength
λ_0	Bragg wavelength of the unloaded state
ζ	Transition

Chapter 1
Introduction

Why is sustainable aviation important?

It is the responsibility of every generation to ensure sustainable development. This means a "development that meets the needs of the present without compromising the ability of future generations to meet their needs" [Wor87, Chap. 2, first paragraph]. The anthropogenic climate change is the most common indicator that human development has not been sustainable recently.

The aviation sector is no exception. Müller et al. [MWW22, pp. 16–17] illustrate that this sector alone is responsible for 2.5 % of the global CO_2 emissions and point out that the occurrence of those emissions at high altitudes further increases their global warming potential. Considering time being a critical factor to avoid irreversible damage to the ecosystem, the authors emphasise the urgent need to reduce greenhouse gas emissions before 2050 [MWW22, pp. 10–11]. The authors thus conclude:

> The question is not whether aviation should take action on sustainability, but rather what does sustainability transformation for aviation entail and how quickly can the aviation industry transform towards sustainability? [MWW22, p. 2]

In accordance, the European Commision [Eur20, p. 2] set the goal to reduce the emissions in the entire transport sector by 90 % by 2050, which includes the objective of zero-emission aircraft becoming market ready by 2035.

Making aviation sustainable, however, is not only a matter of ensuring the ecological system is not put at risk. Sustainability usually includes a relationship between

J. Steinmetz, *Functional Compliance of Multifunctional Bondlines with Disbond Arrest and Health Monitoring Features for Safe and Efficient Composite Joints*, Mechanics and Adaptronics, https://doi.org/10.1007/978-3-031-72422-0_1

environmental, societal, and ecological aspects [GHO02, p. 188]. The aviation sector illustrates what that means.

Aviation has become an integral part of modern societies since the emergence of commercial flight [MWW22, p. 2]. It has brought the freedom to travel almost anywhere on the globe at little effort and expense for people in industrialised and an increasing number of developing countries. On the one hand, this has been accompanied by a boost to the global economy [MWW22, p. 7]. The economic benefits and the new mobility also increase the living standards of many people and improve the multicultural collaboration by connecting people from different countries. On the other hand, these positive effects stand in contrast to the aviation sector being a major driver of the anthropogenic climate change in its current form due to its dependency on fossil fuels [MWW22, p. 2]. Thus, making aviation sustainable means maintaining the socio-economic benefits without putting the stability of the ecological system at risk. This is why sustainable aviation is important.

Technological advances regarding energy efficiency are crucial for reaching the goal of sustainable aviation. According to Torenbeek [Tor13, pp. 51–52], energy efficiency is defined as the available seat-kilometres per litre fuel consumed and mainly depends on four parameters. These are the calorific value of the fuel consumed, the overall engine efficiency, the aerodynamic efficiency (lift-to-drag ratio) and the structural efficiency (payload-to-gross-weight ratio). The fact that reducing fuel consumption is not only of high ecological but also economic interest illustrates the key role an increased energy efficiency plays for sustainable aviation.

Even though the ecological impact of aircraft depends a lot on technology, the technology is only one of five interacting subsystems that influence its sustainability [MWW22, pp. 6–8]. These interactions lead to conflicts. For example, an important limitation for technological innovation comes from the political/regulatory subsystem in the form of safety regulations, which hinder the innovation process [MWW22, pp. 7]. This area of conflict between technological advances and safety is of great importance in this thesis.

The approach of a *multifunctional bondline* presented in this thesis aims at improving the sustainability of composite aircraft by more efficient joints. As explained in detail in Sect. 2.1, structural bonding is a key technology for further increasing the efficiency of aircraft after the development from metallic to composite structures. Safety concerns prevent the use of structural bonding without additional fasteners at present, which decreases the efficiency of composite aircraft due to weight penalties. The main objective of the multifunctional bondline approach is to provide an efficient joining technology for more sustainable aircraft that satisfies the high safety demands for aviation systems. For that matter, the safety of lightweight structural bonds shall be increased by adding disbond arrest and Structural Health Monitoring (SHM) functionalities. Since the safety functions might add weight to the system and could cause disturbances to each other and the bondline, the central research question of this thesis is how these functions can be implemented in compliance with each other and the objective of an increased aircraft sustainability. This is the question of *how to achieve a functionally compliant solution for a multifunctional bondline.*

Chapter 2
Novel Multifunctional Bondline for Efficient and Safe Joints in Composite Aircraft

How could multifunctional bondlines improve the efficiency of a composite aircraft?

This chapter illustrates how the novel multifunctional bondline promises an increased structural efficiency by providing a safe and efficient joining technology for composite aircraft. First, it shows why joining technologies are important for further improving the structural efficiency of aircraft and what the shortcomings of state-of-the-art joining technologies are. Then, it discusses current research approaches to overcome these shortcomings and introduces the novel multifunctional bondline, including the main research need for its successful implementation and the objectives of this thesis. It follows a brief overview of the main idea for obtaining a functionally compliant multifunctional bondline and the investigation methods of this thesis. The chapter closes with the outline of this thesis.

2.1 Lightweight Potential and Safety Concerns of Structural Bonding

What makes structural bonding the favourable joining technology for lightweight structures, and what prevents exploiting its full potential?

Joint design and manufacture are a major challenge for safe, cost-effective, and efficient aircraft structures and can determine the structural efficiency [BS16, p. 397]. Niu [Niu10, p. 285] points out that this is because joints are among the most common sources of failure in aircraft structures due to many factors like secondary stresses, stress concentrations and excessive deflections. He further illustrates how this is

J. Steinmetz, *Functional Compliance of Multifunctional Bondlines with Disbond Arrest and Health Monitoring Features for Safe and Efficient Composite Joints*, Mechanics and Adaptronics, https://doi.org/10.1007/978-3-031-72422-0_2

an even greater issue for composites than metals because composites are brittle, have very little capacity to redistribute loads, and are incapable of masking design approximations in the way yielding metal does [Niu10, p. 285]. Accordingly, Schmid Fuertes et al. [Sch+15, p. 795] claim "one key technology for the future development of composite aircraft structures is a suitable joining technology."

Typically, mechanical fastening, structural bonding, or a combination of both is used for joining airframe structures [BS16, p. 397]. However, neither technology fully satisfies the demand for a safe, cost-effective, and efficient structure at present.

Mechanical fastening is a mature technology [Niu10, p. 290], but due to the notch sensitivity of composites, it reduces the joint efficiency significantly [Niu10, p. 288]. Even the best fastened joint will most likely not exceed 50 % of the strength of an unnotched laminate [Niu10, p. 299]. Moreover, a huge number of fasteners made from expensive titanium alloys and additional shimming are usually required for composites [BS16, p. 400].

Structural bonding allows for a higher composite joint strength at less weight than mechanical fastening [Niu10, p. 334]. Notice that there does not seem to be a generally accepted definition of the term "structural bonding". In the scope of this thesis, structural bonding is defined according to Hart-Smith [Har18, p. 1236]:

Definition 2.1 *Structural bonding* refers to a type of adhesive bonding for transferring load between members by using epoxy, phenolic, or acrylic adhesives.

The advantages of structural bonding over mechanical fastening mainly originate from the more uniform stress distribution along the bonded area, which enables a higher stiffness and load transmission [dÖA18, p. 2]. Moreover, structural bonding does not require drilling holes in the adherends [dÖA18, p. 2]. Accordingly, Niu [Niu10, p. 288] claims structural bonding can overcome the reduced joint efficiency of mechanical fastening due to the high notch sensitivity of composites. In addition, structural bonding provides a high fatigue strength [dÖA18, p. 2], and can even be more cost-efficient than mechanical fastening because structural bonding does not require shimming [BS16, p. 400].

Schmid Fuertes et al. [Sch+15, p. 795] thus see structural bonding as the more promising joining technology, especially for composite structures, when trading off weight, performance, and cost. Nevertheless, structural bonding is barely used in primary joints of aircraft structures, which are joints whose failure results in catastrophic loss of the aircraft. There are multiple reasons for that, three of which are discussed in the following:

First, the design of structural bonds is often more complex, and there are no simple rules in comparison to mechanical fasteners [dÖA18, p. 3]. This seems related to the fact that there are still knowledge gaps in the fundamentals of structural bonding, like adhesion [Ehr+10] and fracture mechanisms [Nag+22]. In addition, there is a lack of definite mechanical properties from the manufacturers, leading to design uncertainties [EQF11, p. 485].

Second, structural bonds do not retard the spread of damage within the adhesive layer. Consequently, there is a high risk of damage spreading rapidly within a structural bond if the adhesive is the weak link in the joint [Har18, p. 1275].

Last and most importantly, there are major concerns regarding the long-term environmental durability of structural bonds [BS16, p. 405]. This is strongly connected to a series of adhesive failures in bonded aluminium structures of aircraft in the late 1960s and early 1970s, caused by improper surface preparation [Har21, p. 769; BS16, p. 405]. This led to the formation of weak bonds[1]. Ehrhart [Ehr16, p. 20] defines these as bonds whose strength is reduced to 20 % or less in comparison to a healthy bond, and which are not detectable by Non-Destructive Inspection (NDI) methods. This is because the interface between the adhesive and the adherend is either in intimate contact or they are coupled through a liquid [BS16, p. 400]. Weak bonds are not only caused by improper surface preparation and contamination of the bond during manufacturing but also environmental degradation [BS16, p. 400]. Despite the later success of the Primary Adhesively Bonded Structures Technology program [Pot79] in improving the durability of adhesive bonds of aluminium structures significantly by applying appropriate surface preparations, adhesive bonding has been considered as a critical process ever since [Har02, p. 140]. This is why, Hart-Smith [Har21, pp. 763–764] emphasises the need to ensure the adhesive stays stuck to the adherends by implementing sophisticated surface preparation processes.

As a result of these challenges, the trust in structural bonding is low [dÖA18, p. 3], and therefore the certification of pure structural bonding in safety-critical systems like aircraft is very conservative [OJK21, p. 3]. To ensure safety, the European Union Aviation Safety Agency [Eur10, p. 6] and the U.S. Department of Transportation Federal Aviation Administration [US 10, p. 5] define the following certification requirements that have to be met in addition to a qualified bonding process and rigorous quality controls of bonded structures in composite aircraft:

> For any bonded joint, the failure of which would result in catastrophic loss of the aeroplane, the limit load capacity must be substantiated by one of the following methods:
>
> (i) The maximum disbonds of each bonded joint consistent with the capability to withstand the loads [...] must be determined by analysis, tests, or both. Disbonds of each bonded joint greater than this must be prevented by design features; or
>
> (ii) Proof testing must be conducted on each production article that will apply the critical limit design load to each critical bonded joint; or
>
> (iii) Repeatable and reliable non-destructive inspection techniques must be established that ensure the strength of each joint. [Eur10, p. 6]

[1] Weak bonds with intimate contact are often also referred to as kissing bonds. However, the term kissing bonds is not used very precisely in the literature [Ehr16, p. 20].

Since proof testing of every single part is not considered feasible and there is no reliable NDI technology for structural bonds, it is the state-of-the-art to use hybrid joints, which combine structural bonding and mechanical fastening, in primary joints of aircraft [Sch+15, p. 800]. The fasteners are sized to maintain the entire limit load capacity of the joint in case of a global bondline failure. The hybrid joints satisfy the certification requirements by degrading the structural bond to a secondary joint, but the use of mechanical fasteners diminishes the weight, cost, and performance benefits of pure structural bonding [Löb16, p. 26]. The implementation of a joining technology that is suitable for primary joints and maintains the advantages of structural bonding has thus great potential to further improve the structural efficiency of aircraft and other highly loaded, lightweight structures.

One way to overcome the issues of structural bonding is alternative bonding mechanisms, which do not rely on adhesion. In this context, it is the state-of-the-art to establish cross-links between the adherend and adhesive by joining and curing composite parts at the same time. Depending on the state of cure of the adherends, there are three bonding technologies for joining thermoset composites [Sch+15, pp. 796–797]:

1. Secondary bonding: Two fully cured parts are joined via an additional adhesive.
2. Co-bonding: An uncured composite part is cured and joined to a second fully cured part via an additional adhesive in a single step.
3. Co-curing: Two uncured parts are cured and joined via an additional adhesive in a single step.

Secondary bonding and co-bonding do rely on adhesion mechanisms, but co-curing is considered a process, which is based on chemical cross-linking rather than adhesion [Löb16, pp. 9–10]. This means that co-curing processes do not fall under the aforementioned certification regulations, and thus co-curing is not considered an adhesive bonding technique even though a film adhesive is commonly used [Löb16, p. 9]. However, Hart-Smith [Har21, p. 797] points out that co-curing is much pricier than secondary bonding for high production rates because it needs large tools and more time to establish the bond. It is thus of high interest to find a solution that is also applicable in secondary bonding.

Fusion bonding provides an alternative mechanism for secondary bonding of thermoplastic composites [Löb16, pp. 6–7]. Via fusion bonding, thermoset composites can be welded to thermoplastics as well, but this requires co-curing a thermoplastic film or applying a hybrid interlayer to the thermoset first [AS19, p. 142]. This in turn raises the question of what kind of interface is created between the thermoplastic and thermoset material. Unlike co-curing thermosets to each other, cross-linking cannot be assumed in that case and, thus, an adhesion mechanism applies again. In addition, the challenge that the high temperatures during fusion bonding reduce the strength of the thermoset material has to be overcome [AS19, p. 149]. As a result, fusion bonding of entire structures seems most appropriate for thermoplastic composites.

In conclusion, co-curing and fusion bonding are alternative approaches for safe and lightweight joints of composites, but they are either much less cost-efficient, not applicable to secondary bonding, or limited to specific materials. Hence, it is of great interest to overcome the challenges of structural bonding itself.

Initial Problem Statement

Structural bonding is the favourable technology for efficient joints of aircraft structures. Their relatively uniform stress distribution provides a high stiffness and load transmission, which is suited well for lightweight structures. When joining composites, these advantages are even greater because the composites can remain unnotched.

However, a low tolerance against damage growth of structural adhesives, design uncertainties due to knowledge gaps regarding adhesion mechanisms and fracture mechanics, as well as uncertainties regarding the long-term durability paired with unexpected failures of bonded aircraft in the past are responsible for major safety concerns. Moreover, the manufacturing process is demanding. To overcome the safety concerns, structural bonding is applied with additional mechanical fasteners in primary joints in aircraft today. This is safe but impairs the lightweight potential of structural bonding significantly. Accordingly, the fundamental challenge of this thesis is:

How to obtain safe structural bonds for joining primary aircraft structures without impairing their lightweight potential?

Figure 2.1 summarises the challenges regarding safe and efficient structural bonds.

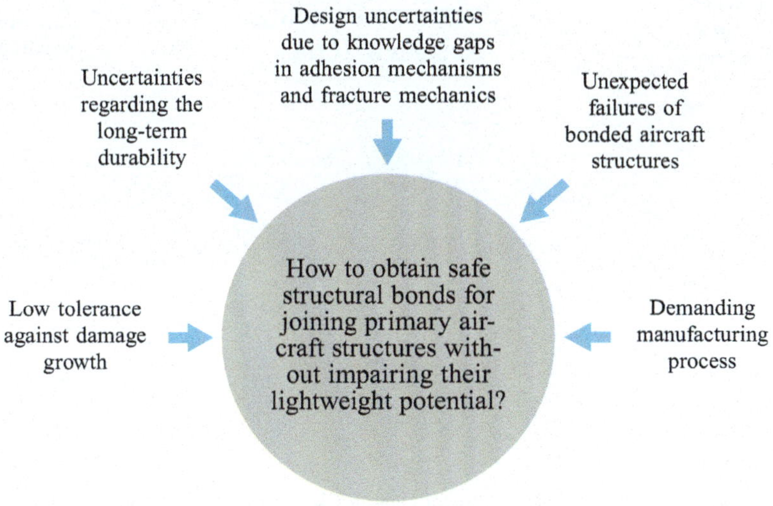

Fig. 2.1 Core challenges for applying pure structural bonding in primary aircraft structures

2.2 Approaches to Safe and Lightweight Structural Bonding

How to obtain safe structural bonds for joining primary aircraft structures without impairing their lightweight potential?

This section discusses approaches to overcome the limitations of structural bonding. It briefly reviews the corresponding research topics and presents the novel multifunctional bondline idea. Figure 2.2 illustrates the discussed research topics and classifies the novel multifunctional bondline approach in that context.

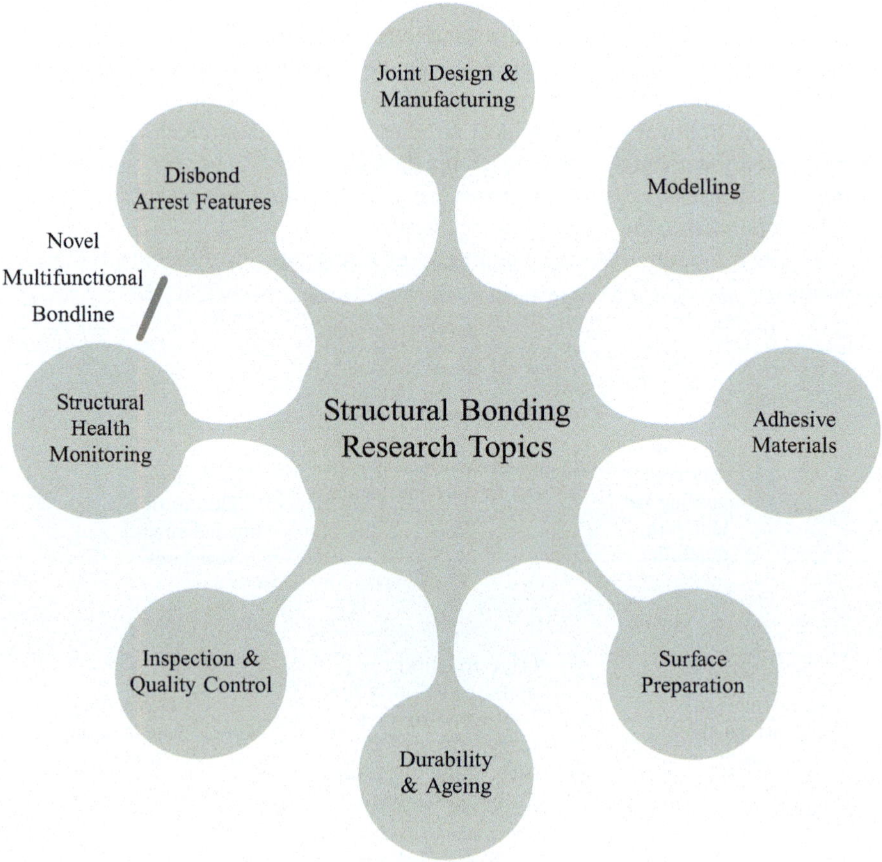

Fig. 2.2 Overview of current research topics regarding structural bonding, and classification of the novel multifunctional bondline approach

2.2.1 Brief Review of the State of Research

What are current research approaches to overcome the safety concerns of structural bonding and exploit its full weight-saving potential?

The potentials and limitations of structural bonding described in Sect. 2.1 have brought a lot of attention to this topic. This section presents current research topics in this field. The goal is not to explain every topic in detail, but to give a compact overview. Hence, the topic selection is based on recent review papers and standard references. Notice that the selection does not claim to be complete, since the topic of adhesive bonding is very broad [Bor+23].

Recalling the challenge of the complex design of structural bonds, there are many efforts to obtain an optimal joint design. Much attention is given to methods how to achieve a more uniform load distribution in overlap joints to fully exploit the load-bearing capabilities of the adhesive and adherends [Bor+23]. In addition, the connection between joint design and environmental degradation, on the one hand, and the fatigue strength, on the other hand, is an important aspect [Ram+20, p. 1307]. In that context, Hart-Smith [Har21, pp. 782–786] shows how a design with a proper load distribution enables durable joints in aircraft that are also resistant to damage in the bondline. The clue is to ensure the adherend fails before the adhesive, in combination with a design that allows a load redistribution in the joint. Damage can not spread in the adhesive then but needs to cause fatigue cracks or delaminations in the adherend to grow. These mechanisms take so much time that the damage is detectable before it reaches a critical size.

Dealing with the complexity and the lack of simple rules, modelling of structural bonds is an important research topic. Numerical models play an increasing role in the selection of a joint's geometry and materials [Bor+23]. For example, Völkerink [Völ22] proposes a method for estimating the failure mode and strength of bonded joints in fibre composite aircraft structures by a progressive damage analysis. There is also much research on modelling the fracture mechanics of structural bonds [Bor+23; Dil21].

Other research focuses on structural adhesives to tackle their low resistance against crack growth. Regarding the adhesive materials, the main goal is to obtain a high strength, high stiffness, and proper ductility at the same time [Bor+23].

A popular alternative approach is the implementation of Disbond Arrest Features (DAFs) into the bondline [ST23; Löb16]. Instead of ensuring damage cannot travel through the bondline at all, the goal of DAFs is to limit the damage to a tolerable size, and therefore increase the damage tolerance of the structural bond. Figure 2.3 illustrates this idea by comparing a DAF with a classical fail-safe mechanism consisting of a bolted joint. The bolted joint provides the limit load capacity of the joint upon failure of the bondline, whereas the DAF prevents the bondline from failing com-

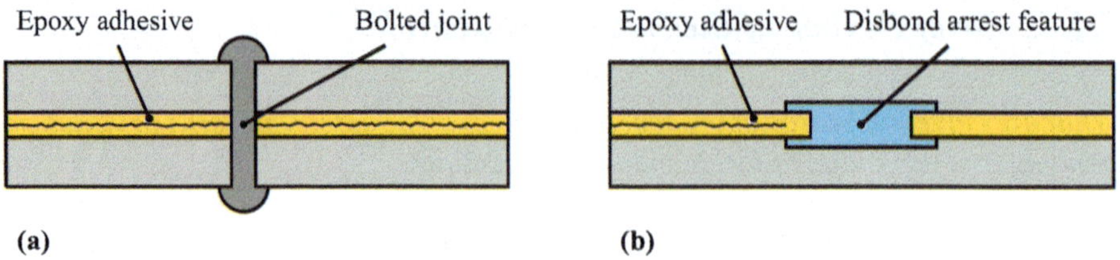

Fig. 2.3 Comparison of fail-safe mechanisms **a** State-of-the-art fail-safe mechanism with an additional bolted joint **b** DAF based on the hybrid bondline approach of [Löb+16]

pletely. The design thus ensures the remaining bondline is still capable of carrying limit loads. Schmid Fuertes et al. [Sch+15] and Löbel [Löb16, p. 25] consider DAFs as the most promising approach to obtain certification of structural bonding in primary structures. Sioutis and Tserpes [ST23, p. 15] agree that the application of DAFs in aerospace structures is feasible but also conclude that more research has to be conducted first. This includes parameter optimisation, the mechanism's design and, especially, fatigue tests.

The environmental durability concerns of structural bonds are the subject of many publications, in addition to the design methods against crack growth [GdC21; Ram+20; Pet15]. For a strong and durable adhesive bond, a sophisticated surface preparation is key [Har21, p. 769]. Hence, there is a great variety of preparation methods for structural bonds [YAL21]. The major issue is that an insufficient surface preparation causes weak bonds, which are not detectable by state-of-the-art NDI methods [BS16, p. 400].

This has led to intensive research in the field of NDI and quality control [Ehr16; Lei+21; CHN21]. In this context, Hart-Smith [Har21, p. 771] demands a mandated durability test of composite structures, which is capable of analysing the quality of the surface preparation at the time of manufacture. In his opinion, there is no point in waiting for a reliable NDI technique if such a test is available [Har21, p. 771]. With the bondline control technology, Heilmann et al. [HWW18] propose a peel test for evaluating the surface preparation of bonded composite repairs. The aim is the certification of critical aircraft structures using a test method, which allows for proof testing every critical joint for severe bonding defects [HH22].

SHM is an alternative, highly regarded research area for increasing the safety of structural bonding [LP22] and composite aircraft in general [BS16, Sect. 13.4]. SHM does not aim at improving the durability of structural bonds directly. Instead, the goal is to access the bondline status during operation and to allow repair or replacement at demand [LP22, p. 2]. Crane et al. [CHN21, p. 245] see SHM as a much more cost-efficient alternative to repeated non-destructive testing at standstill. In addition, Ashcroft [Ash18, pp. 942–943] claims SHM is a promising method to avoid large safety factors when designing structural bonds against fatigue failure. However, he points out the complexity of monitoring fatigue damages as well, leading to the possibility that damage might not be detected early enough if the location is inaccessible or if the critical damage size before rapid fracture is small.

2.2.2 Combining Disbond Arrest with Health Monitoring: The Novel Multifunctional Bondline Approach

What is the multifunctional bondline, how does it contribute to safe and efficient joints, and what is the main challenge towards its application in composite aircraft?

Among the methods presented in Sect. 2.2.1, DAFs and SHM appear very promising for safe and efficient structural bonds. The combination of both leads to the novel *multifunctional bondline* approach. This section briefly discusses the expected benefits and challenges of this approach and names the main objectives of this thesis.

The disbond arrest and health monitoring functionalities complement each other well. The DAF ensures that damage does not rapidly grow to a critical size and hinders the initiation of new damage. Hence, the risk that the SHM feature might not detect damage early enough and the flaw that it does not improve the durability of the bond directly are both overcome. The SHM feature then further improves the safety of the structural bond by detecting damage and allowing the estimation of the residual load capacity of the structural bond during operation. As a result, the first and third options of the aviation authorities for substantiating the limit load capacity of a bonded joint are addressed, simultaneously. This makes the multifunctional bondline approach very attractive to achieve certification for bonded primary structures in the aviation industry. In addition to obtaining certification of bonded primary structures, the multifunctional bondline could further improve the lightweight potential of bonded joints. As mentioned in Sect. 2.2.1, the SHM system could enable a joint design with reduced safety factors. Hence, similar limit loads could be substantiated with a lighter design.

The possibilities the multifunctional bondline offers come with risks, too. While functions may complement each other, they could also disturb one another. In fact, the complementing nature of functions does not even mean that their implementation is complementary as well. The components of the SHM system could cause stress concentrations within the bondline and other structural parts, reducing their load capacity below the design requirements, for example. This means that the functionalities of the multifunctional bondline might not only disturb each other but also affect other functions of the system they are a part of.

The successful implementation of multifunctional bondlines depends on keeping the disturbances at a level where all requirements of the system can be fulfilled. For that purpose, Sinapius [Sin21, p. 2] proposes a concept called *functional compliance*[2] to ensure the requirements of all functions are harmonised in the design phase. (See

[2] Functional compliance is often also referred to as functional conformity. Within this thesis, the term "functional compliance" is used because it is closer to the corresponding adjective.

Sect. 3.1.1 for the exact definition of functional compliance.) Finding a functionally compliant design is considered the main challenge for the successful implementation of multifunctional bondlines.

The Multifunctional Bondline Approach

The multifunctional bondline is a promising approach for safe and lightweight joints. It uses the synergistic effects of disbond arrest and SHM features not only to meet the requirements for bonding primary structures but also to further increase the efficiency of the bondline design. The main challenge for the multifunctional bondline at this point is to prove that a functionally compliant implementation is feasible. The central research question of this thesis is thus: *How to achieve a functionally compliant solution for a multifunctional bondline?*

A literature review on the combination of the disbond arrest and SHM functions for safe and efficient structural bonds confirms the novelty of the proposed multifunctional bondline approach. The method is documented in appendix A.

Only Kadlec et al. [KRB20] propose a system combining disbond arrest and SHM for structural bonds of composites, with the aim of certification of structural bonding in the aviation industry as well. In addition, they share the vision to reduce load factors or increase the limit load of the bonded structure via this combination. They use a z-pinning approach for crack arrest, which means reinforcing the joint locally in the through-thickness direction with small metallic pins. Kadlec et al. [KRB20] study the feasibility of measuring the resistance of the z-pins in crack growth experiments for SHM purposes. According to their measurements, the resistance change of the z-pins correlates with the fatigue crack growth in the bondline [KRB20, Fig. 8]. They explain the increase in resistance with an increase in damaged pins due to the crack growth. Hence, they suggest measuring the resistance of the pins continuously to monitor the bondline integrity.

Kadlec et al. [KRB20] do not investigate the main challenge regarding a functionally compliant design of multifunctional bondlines. They neither discuss the requirements for an actual lightweight design nor a concept to measure the resistance of the pins in an aircraft. Moreover, the z-pinning approach for disbond arrest is not applicable to secondary bonding [Löb16, p. 34]. This illustrates the novelty of the multifunctional bondline approach to safe and efficient joints of composite aircraft.

Objectives of the Thesis

Two objectives of this thesis follow from the novel multifunctional bondline approach. The main objective is to investigate the feasibility of combining the structural bonding, disbond arrest, and SHM functions in a functionally compliant manner for increasing the sustainability of aircraft in the future. The second objective is to investigate whether the findings regarding the multifunctional bondline approach provide useful extensions to the functional compliance concept in general.

2.3 Design and Investigation Methods for a Functionally Compliant Multifunctional Bondline

What are the methods in this thesis for studying the functional compliance of the presented multifunctional bondline approach?

This section describes the approach and research method regarding the central research question of how to obtain a functionally compliant multifunctional bondline. Detailed descriptions follow in Chaps. 3 and 4.1, respectively.

Approach to a Functionally Compliant Multifunctional Bondline

The meaning of **functional compliance** depends on the chosen system boundaries. Within the system of an entire aircraft, functional compliance means meeting the requirements for a safe operation of the bonded structures, while contributing to the sustainability throughout all life cycle phases of the aircraft. Within the subsystem of an overlap joint it means: harmonising the structural bonding, disbond arrest, and SHM functions so that the limit load capacity is substantiated at all times during operation, the multifunctional bondline is manufacturable, and the bonded structure is not heavier than a similar one joined with state-of-the-art joining technologies.

Investigating the functional compliance in the system boundaries of the overlap joint is the focus of this thesis. In this context, the new aspect of managing the system complexity for obtaining functional compliance is proposed. In the scope of the multifunctional bondline, this means keeping the system complexity as low as possible despite adding the disbond arrest and SHM functionalities by reducing the requirements of complementing functions.

The **approach** to a functionally compliant multifunctional bondline is the implementation of Multifunctional Disbond Arrest Features (MDAF). As illustrated in Fig. 2.4, these consist of two stages of Poly(vinyliene fluoride) (PVDF) strips for disbond arrest and integrated strain sensors for SHM. The idea behind this approach is that the epoxy adhesive is only disturbed locally to ensure the multifunctional

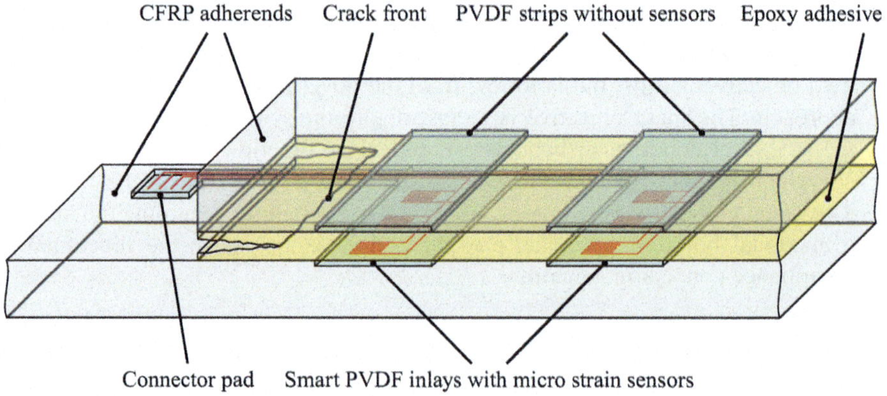

Fig. 2.4 Multifunctional bondline with a two-stage MDAF based on smart inlays with micro strain sensors in a CFRP structure

bondline has similar properties as a state-of-the-art epoxy bondline. This concept uses the proposed synergetic effect of the disbond arrest and SHM functions by arresting the disbond near the sensors to simplify the damage identification. The damage detection itself relies on a longitudinal strain gradient near the crack front. This leads to similar strain outputs in all sensors if there is no crack. However, if there is a crack close to the sensors, the sensors near the crack show a higher strain than those further away, which provides the information to identify the crack.

Investigation Methods

The core investigation method regarding the functional compliance of the MDAF is a **crack growth test** of Cracked Lap Shear (CLS) specimens with different MDAF configurations. This simulates the operation of a bondline with MDAF. Figure 2.4 sketches a CLS specimen, which is a doubler joint with a single free overlap edge and a 10 mm artificial disbond at the overlap edge. The specimens are subjected to a tension-tension cyclic load over 10^6 load cycles. Different specimens are tested at three load maxima, which correspond to a typical limit strain and ultimate strain level of Carbon Fibre Reinforced Plastic (CFRP) and an interim load level. The advantage of that method is that it allows for the simultaneous evaluation of the dynamic strength of the bondline and the functionality of the disbond arrest and SHM features. This is crucial for studying their functional compliance. For the evaluation, the crack growth is tracked by an optical measurement at standstill, which means the test is interrupted once a minute or every 480 cycles. Applying a speckle pattern to one side of the CLS specimens also allows for measuring the strain using Digital Image Correlation (DIC).

Besides the MDAF with the PVDF-based smart inlay depicted in Fig. 2.4, the investigations include two variants with Polyetherimide (PEI)-based smart inlays and PVDF cover, and an MDAF made of PVDF strips and Fibre Bragg Grating (FBG) sensors. One of the PEI-based smart inlay variants is made for investigating whether it is beneficial to place the connector pad on the other side of the CLS specimen to keep the track bridges away from the crack growth area. CLS specimens with sensorless DAFs made of PVDF or PEI and CLS specimens without DAF complete the selection. This allows for assessing the influence of the integrated MDAFs and DAFs on the bondline strength.

Quasi static tests until failure, including new CLS specimens and CLS specimens after the crack growth test, complete the investigations. These shall provide information on the influence of the integrated MDAFs on the static strength of the structural bond. Moreover, some new CLS specimens are equipped with an increased artificial disbond length for obtaining a clearer correlation between the sensor outputs and the crack length. Furthermore, some CLS specimens are converted to a kind of Single Lap Shear (SLS) specimen before the quasi-static test to ensure a failure in the bondline. This is achieved by cutting the continuous adherend within the overlap.

The investigations regarding the **manufacturability** of the multifunctional bondline rely mostly on checking the sensor integrity after each manufacturing step of the CLS specimens with MDAF. In the case of the smart inlays, a second method is used for studying how the melting of the PVDF substrate during curing affects the sensor structures. For that matter, a curing process is simulated under a microscope. By comparing uncovered smart inlays with smart inlays placed under a cover glass, the effect of the rigid tooling being in contact with the sensor structures during curing is also simulated. The microscopic images reveal if damage to the sensor structures occurs during curing, and the cause of damage.

Since PVDF can be a piezoelectric material depending on its phase [MLL14, p. 684], additional investigations deal with the feasibility of utilising the piezoelectric effect for the sensing function of an MDAF after curing. For this study, the phase of polarised PVDF foils is determined before and after annealing in an autoclave at various maximum temperatures and 3 bar pressure. The amount of piezoelectric phase after curing determines whether it is feasible to utilise the piezoelectric effect.

The **evaluation** of the disbond arrest effect is based on the crack tips in the crack growth tests and an analysis of which interface is responsible for the failure if a crack has passed the MDAF. The SHM function and sensor integrity are studied based on the correlation between the crack growth measurement and the output signals. Moreover, the images of the bondline damages are considered for evaluating the cause of sensor failures if they occur. Correlating both results also provides information regarding the proposed complementing nature of the disbond arrest and SHM functions.

A functionally compliant operation of the MDAFs implies a successful integration of the disbond arrest and SHM features during manufacturing. The sensor structures in particular might be damaged during manufacturing. Hence, the functional compliance investigation includes the manufacturing process as well.

2.4 Thesis Outline

This thesis investigates the question of how to achieve functional compliance for a multifunctional bondline on three nested levels. The top level is the general concept of functional compliance. The second level is the functional compliance of a multifunctional bondline of aircraft, and the third level is the functional compliance of an overlap joint with a multifunctional bondline. The third level builds the core of the three levels and is the most relevant in this thesis. In the following, the main chapters are briefly explained regarding that structure. The outline is depicted in Fig. 2.5.

Chapter 3 defines the three levels and builds the foundations for all investigations first. For that matter, it explains the functional compliance concept in general for the top level and derives the meanings of functional compliance for a multifunctional bondline regarding the system boundaries of an aircraft and the system boundaries of an overlap joint for the two nested levels. In the following, Chap. 3 focuses on the level of the overlap joint to provide the framework for the experimental investigations regarding the feasibility of combining the structural bonding, disbond arrest and SHM functions in a functionally compliant manner. This includes a literature review on DAFs and SHM approaches for structural bonding, the definition of the research hypotheses and proposing the main working principles and two initial design concepts for a functionally compliant multifunctional bondline utilising MDAFs.

Chapter 4 deals with the experimental studies for testing the functional compliance of the proposed bondline with MDAFs on the level of the overlap joint. It therefore presents the investigation methods and specimen design for testing the research hypotheses first. This includes a total of four alternative MDAF designs. The core of Chap. 4 is the experimental results and their discussion. Based on these, the chapter closes with an evaluation of the functional compliance of the different MDAFs.

Chapter 5 takes the experimental results on to the level of the multifunctional bondline in aircraft. Therefore, it discusses the findings and further challenges regarding the functionally compliant implementation of the multifunctional bondline in the fuselage of a composite aircraft through fundamental questions from Chap. 3 and two applications scenarios for the fuselage of composite aircraft.

Chapter 6 takes on the results from the previous chapters to the top level by discussing what the findings contribute to the general concept of functional compliance. For that matter, it deals with additional aspects of functional compliance and the core challenges regarding the evaluation of functional compliance that have been encountered within the investigations of functionally compliant multifunctional bondlines.

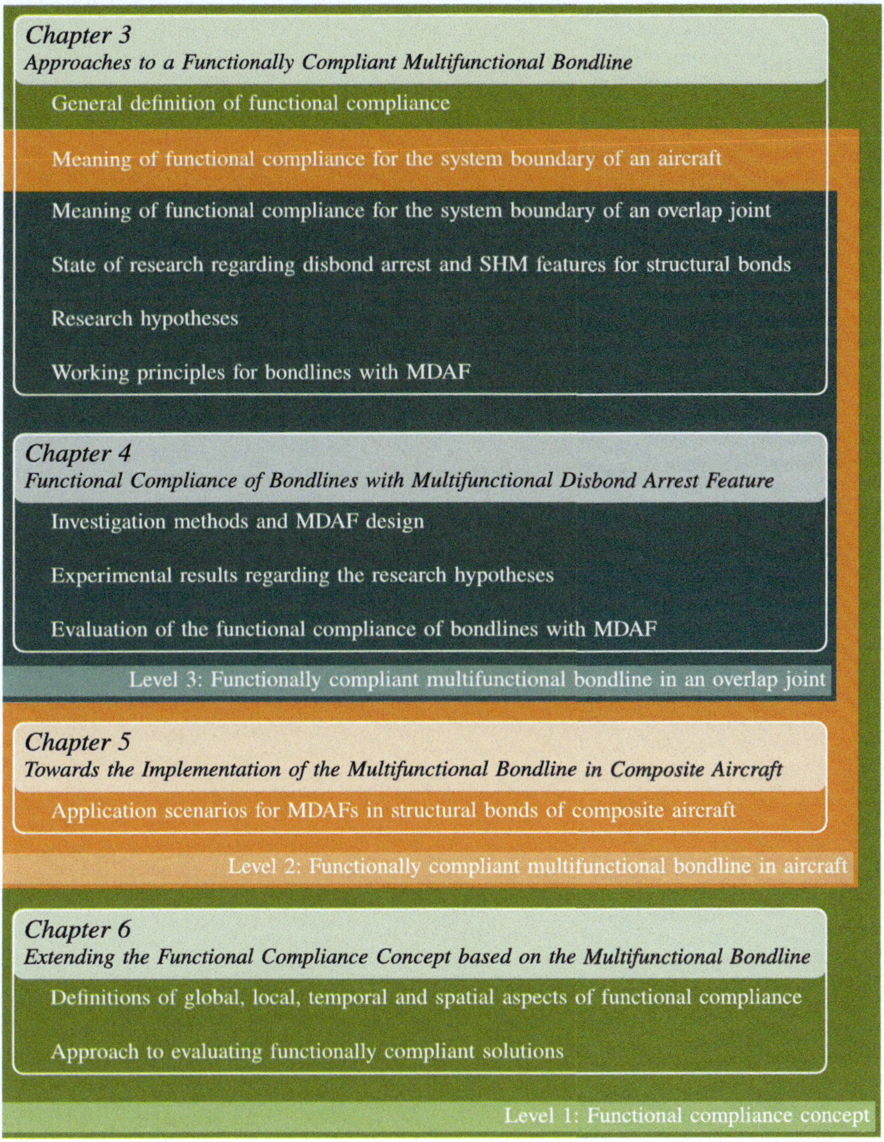

Fig. 2.5 Graphical outline of the thesis showing the main chapters with the core results assigned to the three levels of functional compliance

Chapter 3
Approaches to a Functionally Compliant Multifunctional Bondline

Central Research Question
How to achieve a functionally compliant solution for a multifunctional bondline?

Finding an answer to the central research question of how to achieve a functionally compliant solution for a multifunctional bondline requires a thorough understanding of how the functions of the multifunctional bondline interact with each other and the entire system they are a part of. Hence, this chapter discusses why requirements from all phases of a system's life cycle have to be considered to ensure a functionally compliant solution. From this broad view, the focus is directed towards more fundamental questions regarding the working principles of multifunctional bondlines by narrowing down the system boundaries to an overlap joint and introducing three subsequent research questions. The goal behind this is to obtain a better knowledge of more elemental aspects of the novel multifunctional bondline approach before investigating the very complex interactions in a complete system.

For studying the functional compliance regarding the working principles, this chapter presents a research hypothesis for each research question and develops a fundamental design approach. Both of these follow from the definition of fundamental requirements for the multifunctional bondline in composite aircraft and a thorough literature review. Parts of this chapter have been published previously in [Ste+21].

J. Steinmetz, *Functional Compliance of Multifunctional Bondlines with Disbond Arrest and Health Monitoring Features for Safe and Efficient Composite Joints*, Mechanics and Adaptronics, https://doi.org/10.1007/978-3-031-72422-0_3

3.1 Meaning of Functional Compliance for the Multifunctional Bondline

What does functional compliance mean for the multifunctional bondline in composite aircraft?

This section introduces the concept of functional compliance in general and the multifunctional bondline in particular. The section shows the importance of considering the application when defining the meaning of functional compliance. Accordingly, the application scenario of a structurally bonded fuselage of composite aircraft serves for formulating fundamental questions and requirements for multifunctional bondlines. These are further refined by introducing two different system boundaries: a composite aircraft and an overlap joint. This yields two meanings of functional compliance for the multifunctional bondline.

3.1.1 Functional Compliance

What is functional compliance?

Sinapius [Sin21, p. 2] defines functional compliance as "optimum of the competing and non-competing objectives for all functions in an adaptive structure". Even though he explicitly names adaptive structures, he justifies the demand for this concept by the multifunctionality of adaptive structures [Sin21, p. 2]. Hence, functional compliance can be understood as applying to all multifunctional systems, and thus to the multifunctional bondline.

According to a discussion with Sinapius, his definition of functional compliance refers to an optimisation goal in accordance with the VDI 2221 [Ver19]. For a more profound understanding of the meaning of functional compliance, the VDI 2221 is thus analysed briefly.

The VDI 2221 describes the design process as an iterative process, which takes place in "procedural principles" such as "analysis/synthesis, abstraction/concretisation and variation/restriction" [Ver19, pp. 19–21]. Furthermore, it proposes a model for product design with nine activities to support the product development process [Ver19, pp. 35–42]:

1. Clarifying and itemising the problem or task
2. Determining functions and their structures
3. Searching for solution principles and their structures
4. Assessing and selecting the solution concept
5. Subdivision into modules – interface definitions
6. Designing the modules

7. Integrating the product as a whole
8. Elaborating the details of execution and use
9. Assurance of the fulfilment of the requirements

There is no explicit formulation of the actual optimisation target in the VDI 2221. However, the ninth activity, which includes analysing if a solution is suitable for all stakeholders, implies that the design optimum is the solution that satisfies the stakeholders most. "Functional compliance" can thus be interpreted as describing the optimal solution for a multifunctional system regarding the competing and non-competing objectives of all functions coming from all stakeholders. Consequently, in this interpretation, there is only one functionally compliant solution, while all other solutions are not considered functionally compliant.

The interpretation of functional compliance as the optimal solution is very attractive for the later stages of product design when the selection of the optimal solution is crucial. Gericke et al. [Ger+21a, p. 68, Table 4.1] support this by explicitly including "optimisation" and "selection of the solution" in their description of the ninth activity of the VDI 2221. However, the VDI 2221 emphasises that obtaining "better solutions and optimisations" does not involve all nine activities being carried out strictly sequentially, but iteratively [Ver19, p. 35]. Hence, the process of getting to the functionally compliant solution is not a single optimisation but involves all nine activities. In addition, a closer look at how Sinapius [Sin21, p. 2] derives his definition reveals that the basic concept of functional compliance is not only about the optimisation of a system with competing and non-competing functions but also about the harmonisation of all functions in the design phase. Hence, the following discusses how the basic concept of functional compliance could be applied besides optimisation tasks in the design process, and how this comes with further interpretations of "functional compliance" than describing the optimal solution.

The second and third activities of the VDI 2221 deal with the definition of functions and groups of functions and the search for solution principles and their structures [Ver19, pp. 36–39]. At this stage, the goal is not to find the optimal solution but to obtain a wide search space of different solution principles and their structure. The VDI 2221 states that this approach to widening the search space is particularly useful for "new developments or fundamental problem solutions" [Ver19, p. 36]. The VDI 2221 further suggests identifying the solution principles and their structures by finding effective principles that are solutions for the main partial functions first [Ver19, p. 37] and linking those effective principles to solution principles for the overall function afterwards [Ver19, p. 38]. The results are fundamental solutions that map the effective structure for fulfilling individual functions or the entire function structure of the product. The VDI 2221 explicitly mentions how the third activity includes the identification of other functions that enable or prevent specific solution principles [Ver19, p. 38]. Hence, the basic concept of functional compliance to harmonise all functions of a system seems appropriate for identifying the fundamental solutions, which describe the entire function structure. However, the result at this stage should be all fundamental solutions and not only a single, potentially optimal one. This is why the second interpretation considers an effective structure "functionally compliant" as soon as all objectives of the corresponding system are fulfilled. In this interpretation, functional compliance is a threshold that distinguishes if an effective structure is a fundamental solution for a given task or not. This interpretation is not limited to effective structures, but also applies to designs of subsystems, as discussed in the explanation of definition 3.1.

The meaningfulness to distinguish between interpreting functional compliance as threshold or optimum depending on the stage of the product development process is supported by Ziebart [Zie12, pp. 64–66], who points out that decisions at early design stages have to be made at a rather high level of uncertainty. This means that at the early design stage, it is often uncertain which fundamental solution is the optimal one.

On the contrary, it is usually impossible to create a detailed design for all fundamental solutions [Ver19, p. 38]. Hence, the VDI 2221 proposes to assess and select the most promising fundamental solutions by suitable evaluation criteria as another activity [Ver19, p. 38]. To describe the selection related to the concept of functional compliance, it is useful to interpret functional compliance as a measure rather than as an optimum or threshold. The measure ranks the solutions according to the defined evaluation tools. One solution may thus be described as more or less functionally compliant than another one. The top-ranked solution would be the optimal one for the given state of the design process. However, depending on the stage of the design process the evaluation is not final, and thus it is not clear whether the top-ranked solution will be the optimum at the end of the design process as well. Nevertheless, the need for ranking solutions explains the third interpretation of functional compliance being a measure to describe the value of alternative solutions.

The analysis of the design process regarding the VDI 2221 shows that all three interpretations of functional compliance are important for different tasks in the design process. However, one term cannot possibly describe all solutions, the value of the solutions, and the optimal one at the same time, as these are three related yet different things. Hence, this thesis uses the following definition to provide a clear distinction between all three interpretations:

Definition 3.1 *Functional compliance* is a *threshold* that describes whether a design fulfils the demand of harmonising all functions in a multifunctional system upon functional integration, regarding all requirements of the system. The *functional compliance score* is a measure that further describes the value of all functionally compliant designs by specific evaluation methods. The maximal functional compliance score defines the *functional compliance optimum* for given requirements and evaluation methods.

In this definition, the core meaning of "functional compliance" is shifted from the interpretation as an optimum to the interpretation as a threshold. The interpretation as an optimum is not discarded, but expressed as "functional compliance optimum" instead. The reason behind this is that the term "functional compliance" includes all designs that are a solution for the given task. Accordingly, the optimal functionally compliant design is also included in the set of functionally compliant designs. If functional compliance addressed the optimal solution only, a new set would have to be created for all other designs that are a solution but not the optimal one. This appears to be counterintuitive and is thus avoided.

In addition, the definition speaks of multifunctional systems, which include functional integration, instead of adaptive structures. This addresses the fact that the creation of functional compliance is not only of concern for adaptive structures but all multifunctional systems that consist of integrated functions.

The common practice of designing complex products by defining subsystems is another important reason to use the system term in the definition [Ver19, p. 39]. As a result, the design concept to create functional compliance applies directly to subsystems as well and is therefore much dependent on the system boundaries. The impact of this will be discussed in detail when analysing the meaning of functional compliance for the multifunctional bondline in the following sections and in Sect. 6.1.

Instead of objectives for each function, the requirements of the multifunctional system are considered in Definition 3.1. This is because requirements are normally used for precisely setting the boundaries for product design in engineering. Moreover, requirements are often provided as minimal requirements and include wishes. Hence, in the early design stage, functional compliance could be ensured by developing designs that fulfil all requirements. Then, the functional compliance scores could be obtained directly from the minimal requirements and wishes by analysing which solutions exceed minimal requirements and meet wishes. Weighted requirements could be applied as an evaluation method in that case, for instance. In the later stages, the best-evaluated solution forms the functional compliance optimum regarding the defined requirements and evaluation method for the evaluated system.

Potential benefits of this definition are illustrated by the example of the multifunctional bondline in this thesis and further discussed in Sect. 6.1.

3.1.2 *Initial Concept for Multifunctional Bondlines in the Fuselage of Composite Aircraft*

What are fundamental questions regarding the implementation of multifunctional bondlines in composite aircraft?

This section presents an initial concept for implementing multifunctional bondlines in primary and secondary joints of the fuselage of composite aircraft. For the initial concept, an existing application scenario from Löbel [Löb16, pp. 179–181] for structural bonds containing thermoplastic DAFs serves as the baseline. This section then analyses what main questions arise when expanding this concept to a multifunctional one. The aim is providing the fundamentals to discuss the main challenges towards a functionally compliant multifunctional bondline in composite aircraft, qualitatively. A rough quantification of specific design parameters like the distance between DAFs is discussed later in Chap. 5.

Löbel [Löb16, pp. 179–181] proposes the longitudinal or circumferential splices and stiffener elements of a fuselage as application scenarios for structural bonds with DAFs in primary joints. At present, mechanical fasteners are used for these joints. Moreover, the stiffeners and frames of the fuselage are an attractive application for structural bonding of secondary structures [Löb16, p. 179]. Hence, the fuselage is well suited to analyse the potential and challenges for the application of multifunctional bondlines.

Figure 3.1a illustrates the initial concept of a fuselage section with stiffeners and stringers structurally bonded to the skin. In addition, positions for circumferential and longitudinal splices of fuselage sections are marked. Thermoplastic DAFs are placed periodically to secure the structural bonds. This arrangement is only shown for the bonds of the stiffeners and frames to the skin, but a similar principle applies to the splices as well [Löb16, pp. 179–180].

There are two types of joints in the concept: a doubler configuration for the stiffeners and frames, and a wide overlap configuration for the splices [Löb16, p. 180]. Figure 3.1b shows the doubler configuration with DAFs. The DAFs are thermoplastic barriers, which are embedded into the surface layer of both adherends. In combination with the epoxy adhesive, the thermoplastic DAFs form the hybrid bondline developed by Löbel et al. [Löb+16]. The hybrid bondline is also explained in more detail in Sect. 3.2.1.

With structural bonding and DAFs, the initial concept already includes two of the three functionalities of the multifunctional bondline. The main challenge is thus to extend the initial concept by an SHM system. However, the DAF itself is still in the conceptual phase, and it has yet to be proven that it meets all requirements for the application in a composite fuselage. The design of the DAF is thus also an important part of the investigations regarding the multifunctional bondline concept for composite aircraft.

Extending the initial concept towards a multifunctional one directly raises many questions regarding the SHM system. It has to be clarified what properties shall be measured, how to distribute the sensors, how to supply the sensors with energy, and how to transmit and evaluate the sensor signals. All this has to be achieved in combination with the structural bonding and disbond arrest functionalities. This illustrates why an important question regarding functional compliance is how to combine these functionalities in such a way that they neither disturb each other nor other functions of the aircraft. This does not only include the operation of these functions but also requires a stable manufacturing process, ensuring the integrity of all the multifunctional bondline's components. Finally, the multifunctional bondline

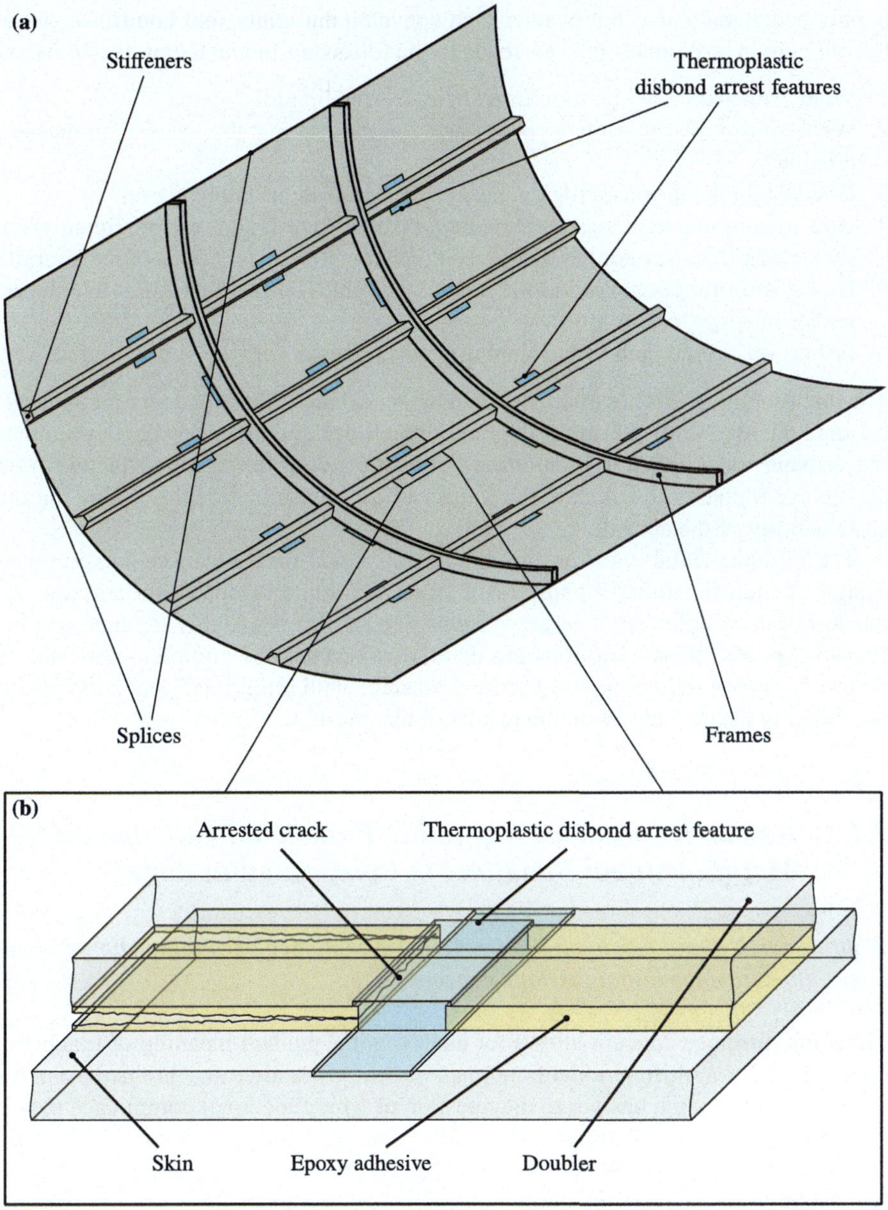

Fig. 3.1 Initial concept for structural bonding with DAFs in a fuselage of composite aircraft **a** Fuselage section with positioning of stiffeners and frames with DAFs, and splice positions **b** Detailed view of a doubler configuration with thermoplastic DAF for joining the stiffeners and frames with the fuselage skin

is only beneficial if it is not heavier than conventional joints, and contributes to an overall gain in sustainability. This leads to the following fundamental questions:

1. What properties shall be monitored to assess the bondline status?
2. What sensor distribution is necessary for monitoring the chosen properties, reliably?
3. How to supply all sensors with energy and access their output signal?
4. How to combine the structural bonding, disbond arrest, and sensing functions in such a way that they neither disturb each other nor other functions of the aircraft?
5. Is the multifunctional bondline manufacturable? (Manufacturing tolerances, sensor integrity, etc.)
6. Is there an overall gain in sustainability when considering the entire product life?

The first four questions are mainly related to the technical challenges for the successful implementation of the multifunctional bondline. The last one not only implies the demand for a lightweight solution but also for considering other factors from all life cycle phases to evaluate the impact of the multifunctional bondline on the sustainability of the aircraft.

These fundamental questions are the starting point for a functionally compliant design of multifunctional bondlines in structural joints of composite aircraft. As the design of complex engineering systems like aircraft is a highly iterative process [Pah+21, p. 37], these questions are neither expected to be complete nor final. A first revision and follow-up concept for a fuselage with multifunctional bondlines is presented in Chap. 5 based on the results of this thesis.

3.1.3 Research Questions Regarding Functionally Compliant Multifunctional Bondlines in Composite Fuselages

What research questions about the functional compliance of multifunctional bondlines follow from the initial fuselage concept?

The initial fuselage concept allows for analysing the general meaning of functional compliance for multifunctional bondlines in composite aircraft. On this basis, this section presents a first answer to the question of what functional compliance means for the multifunctional bondline. This shows that accomplishing functional compliance is a very complex task, which should be divided into more comprehensible ones for obtaining a functionally compliant design. For that matter, this section presents three subsequent research questions to the central research question and introduces the system boundary of an overlap joint.

The fundamental questions one to four from Sect. 3.1.2 regarding the implementation of a multifunctional bondline in a composite fuselage lead to the first research question: How to combine the disbond arrest, health monitoring, and structural bonding functions for a safe operation of bonded structures, while maintaining the mechanical strength and weight-saving potential of an ordinary, undamaged epoxy bondline? The corresponding investigation of how to aggregate the structural bonding, disbond arrest, and SHM functions for a safe operation of structurally bonded parts is key for evaluating the multifunctional bondline approach. Hence, Sect. 3.3 is about the safe operation of a multifunctional bondline.

Following the fifth fundamental question about the manufacturability of the multifunctional bondline, the second research question asks: Is the multifunctional bondline manufacturable using standard curing processes? Regarding functional compliance, a manufacturing process that ensures the integrity of all components of the multifunctional bondline is as important as the successful combination of the functionalities itself. In addition, the certification processes in the aerospace industry, for example, are costly [Sch+19]. That is why it is of great interest that the multifunctional bondline does not require major changes in the production process, which would require additional certification processes. For this reason, the compliance of the multifunctional bondline with standard curing processes is treated in Sect. 3.4.

The third and final research question selected for this thesis is: How to manage the system complexity of the multifunctional bondline? In contrast to the first two research questions, this one does not come from the fundamental questions. Moreover, it does not appear to directly correlate with the successful implementation of the multifunctional bondline, its proof of concept, or functional compliance in general. At least it is a valid argument, that functional compliance is achieved no matter how complex the subsystems of the multifunctional bondline become, as long as all of them function properly. So why bother thinking about reducing the system complexity in the scope of functionally compliant solutions for the multifunctional bondline?

Complex systems can have both positive and negative consequences. According to Perrow [Per11], on the one hand, complex systems with multifunctional components are desirable as they are often more efficient [Per11, p. 88], and additional functions (or devices) can reduce the chance for some errors in a system [Per11, p. 132]. On the other hand, the same functions also introduce new possibilities of error [Per11, p. 132]. Moreover, the higher the complexity becomes, the greater the risk of an unexpected and unpredictable behaviour, which eventually causes system accidents [Per11, p. 72]. In fact, much of the lack of trust in structural bonding itself originates from such unexpected behaviour of structural bonds of aircraft in the 1960s [Har21, p. 769]. Hence, the system complexity has a direct influence on the functional compliance in terms of the requirements to provide a safe and trustworthy system that is efficient at the same time. In this sense, the multifunctional bondline approach has to satisfy conflicting priorities between an efficient design for structural bonds and keeping the system complexity at a safe level. This is why Sect. 3.5 deals with managing the system complexity of the multifunctional bondline.

The last fundamental question asking for an overall gain in sustainability shows that the answer to the question of what functional compliance means for the multifunctional bondline in composite aircraft goes much deeper than it might appear at first. Clearly, harmonising all requirements for a safe and efficient operation and those for the manufacturing process is a main challenge that has to be met. Otherwise, the multifunctional bondline would simply not work in the first place. Meeting this challenge is a complex task already, but is this sufficient to obtain functional compliance regarding the higher goal of sustainable aviation, which is the core motivation behind the multifunctional bondline concept? Certainly, it is not because only two life cycle phases of the bonded structure are considered. Therefore, the similarly important question regarding the sustainability of the multifunctional bondline in other life cycle phases is neglected.

Taking the requirements of all life cycle phases into account is important for ensuring that the implementation of multifunctional bondlines truly improves the sustainability of aircraft. According to Pinheiro Melo et al. [Pin+20] the life cycle phases of an aircraft are: resource extraction, manufacturing of components, aircraft assembly, aircraft operation, aircraft maintenance, repair and overhaul, and aircraft end of life. If not all of these phases are considered, there is the risk that a gain in sustainability in one life cycle phase is jeopardised by a loss in other phases. For example, the expected additional material input due to the implementation of the multifunctional bondline in aircraft structures might lead to additional emissions during the resource extraction and other phases. This means that even if a reliable solution for a multifunctional bondline is found, which also leads to a lighter structure, this does not guarantee the overall impact on the sustainability of the aircraft is positive. From that perspective, achieving functional compliance for multifunctional bondlines means harmonising the requirements for a safe operation of the bonded structure with those for sustainability across all life cycle phases. Hence, to ensure functional compliance is achieved, a life cycle assessment of a composite aircraft with multifunctional bondlines would be appropriate.

A life cycle assessment requires many iterations [Hau18, p. 60], just like product design itself [Ger+21b, p. 37]. This implies that reaching the overall objective of a functionally compliant solution for the multifunctional bondline within the system of sustainable aircraft is a complex task, which requires many iterations as well. In accordance with the VDI 2221 [Ver19], the focus of this thesis is to provide fundamental concepts for a multifunctional bondline as a first step towards their implementation in aircraft. Hence, the system boundary for discussing the functional compliance in the scope of this thesis is set to the subsystem of an overlap joint with composite adherends and an implemented multifunctional bondline. The importance of considering the requirements of all life cycle phases for a functionally compliant design is briefly discussed in Chap. 5, nevertheless.

3.1.4 Fundamental Requirements and Research Approach to Integrate the Disbond Arrest and SHM Functions into an Epoxy Bondline

What are the fundamental requirements for a functionally compliant multifunctional bondline in composite aircraft, and what research approach follows from these?

This section presents a more specific meaning of functional compliance regarding the system boundary of an overlap joint. For that matter, it analyses the fundamental requirements regarding the combination of the structural bonding, disbond arrest and SHM functions. In addition, this section presents the research approach of this thesis.

Table 3.1 lists the fundamental requirements for the multifunctional bondline. They address the core demands of a safe and lightweight joining technology from the certification regulations for aircraft described in Sect. 2.1. Moreover, specific requirements regarding the combination of the disbond arrest, SHM and adhesive bonding functions are included in this list for considering the three central research questions from Sect. 3.1.3. In summary, the requirements demand maintaining the mechanical properties of an undamaged epoxy bondline, increasing its damage tolerance, and monitoring its status, while adding as little weight to the bonded structure as possible, and being compatible with standard curing processes. Note that there are no additional fundamental requirements stemming from the third research question about complexity reduction, because the existing requirements already imply a suitable system complexity. For example, requirement 1.4, demanding an unambiguous distinction between a damaged and a healthy bondline, implies that false alarms due to a too-high or too-low system complexity are both unacceptable.

Table 3.1 Fundamental requirements for multifunctional bondlines of composite aircraft structures

No.	Requirement
1	The residual strength of the bonded structure must substantiate the design load capacities at all times during operation
1.1	Damages within the bondline must be prevented from rapidly growing to a critical size
1.2	Damages within the bondline must be detected before the integrity of the structure is at risk
1.3	Repeated estimates of the bondline's residual strength during operation must be enabled
1.4	There must be an unambiguous distinction between a damaged and a healthy bondline
2	The total weight of the bonded structure has to be as light as possible. It must not be heavier than a structure joined conventionally, i.e. with additional mechanical fasteners, at least
3	The multifunctional bondline has to be compatible with standard curing processes of composites

Functional compliance for multifunctional bondlines within the system boundaries of an overlap joint means harmonising the structural bonding, disbond arrest, and SHM functions regarding all requirements from Table 3.1. Accordingly, a functionally compliant multifunctional bondline has to substantiate the limit load capacity at all times during operation, has to be manufacturable using standard curing processes, and must not be heavier than state-of-the-art joining technologies.

The fundamental research approach behind the multifunctional bondline in this thesis is to achieve functional compliance upon the integration of the disbond arrest and SHM functions into a standard epoxy bondline. The motivation for the integration of these functions is requirements 1.1 and 1.2, which explicitly ask for actions within the bondline. The idea to integrate the safety features in a standard epoxy bondline accounts for the ability of these to carry the design loads, as long as they are undamaged. For the same reason, Research Question 1 explicitly asks how to reach the mechanical strength of a standard, undamaged epoxy bondline. Together with the three research questions from Sect. 3.1.3 and the meaning of functional compliance for the system boundaries of an overlap joint, the research approach builds the framework for investigating the multifunctional bondline.

Meaning of Functional Compliance for Multifunctional Bondlines
The meaning of functional compliance for multifunctional bondlines depends on the chosen system boundaries. For the system boundaries of composite aircraft, functional compliance means that the multifunctional bondline meets the requirements for the safe operation of the bonded structures, while contributing to the overall sustainability of the aircraft. For the system boundaries of an overlap joint, functional compliance of the multifunctional bondline means harmonising the structural bonding, disbond arrest and SHM functions in a manner it substantiates the limit load capacity at all times during operation, is manufacturable by using standard curing processes, and is not heavier than state-of-the-art joining technologies. The latter is used for the fundamental investigations within this thesis.

3.2 State of Research Regarding Disbond Arrest and SHM Features for Structural Bonds

What is the state of research regarding disbond arrest and SHM features for structural bonding in composite aircraft?

The fundamental approach of integrating the functions of the multifunctional bondline directs the focus towards the need to evaluate whether disturbances between the disbond arrest, SHM, and structural bonding functions occur. If this is the case, the multifunctional bondline design has to ensure that these disturbances are decreased sufficiently to meet all requirements. Hence, a good knowledge of how the disbond arrest, SHM, and structural bonding functions affect each other upon functional integration is necessary for a functionally compliant design. For that matter, research approaches to disbond arrest and SHM features for structural bonds are discussed in this section. The findings from the literature review not only help to define the fundamental working principles of the multifunctional bondline but also emphasise its potential for providing a safe and efficient joining technology.

3.2.1 Review of Disbond Arrest Features for Structural Bonds

What are approaches to DAFs for structural bonds?

There is a great variety of DAFs for structural bonds of composites. Löbel [Löb16, Chap. 4] and Sioutis and Tserpes [ST23] both present profound reviews on this topic. This section provides an overview of DAFs based on their evaluation. The most promising methods regarding the multifunctional bondline are then further analysed. However, the term "disbond arrest feature" is clarified first.

Clarification of the Term Disbond Arrest Feature

"Disbond arrest feature" is no standardised term. Consequently, there are some alternative terms found in the literature. For example, Löbel et al. [Löb+16] speak of "disbond stopping feature", and Sioutis and Tserpes [ST23] of "crack arrest feature". Hence, the current terminology is briefly criticised before the DAFs are presented. This requires looking at typical damage types of structural bonds and the wording in the certification requirements because in that context the current terminology can be misleading.

Omairey et al. [OJK21, Fig. 5] illustrate eleven types of defects in structural bonds that all affect their mechanical performance. Disbonds and cracks are only two of them. Disbonds are a separation of the interface between the adherend and adhesive, i.e. an adhesion failure, whereas cracks are a type of cohesion failure within the adhesive or adherend. Does that mean disbond and crack arrest features are actually two different things?

The European Union Aviation Safety Agency [Eur10, pp. 31–32] and U.S. Department of Transportation Federal Aviation Administration [US 10, p. A2-2] follow this definition of disbond, and in their certification rules they explicitly demand the size of disbonds has to be limited by a design feature (see Sect. 2.1). However, it does not make sense that the certification of structural bonding only requires ensuring one type of defect does not reduce the limit load capacity below the design value. Hence, it has to be assumed that in the certification rules, disbond refers to adhesion and cohesion failures alike, and thus covers any kind of growing defect in the structural bond, which decreases its strength. Therefore, the terms "crack arrest feature" and "disbond arrest feature" are considered synonyms, even though, strictly speaking, they refer to different defect types. This is supported by the fact that regardless of whether called crack or disbond arrest, the features in the literature aim at arresting every kind of growing defect in the structural bond.

The difference between "disbond stopping" and "disbond arrest" appears less critical. Within this thesis, "disbond arrest feature" is used because this term is in agreement with the wording from the certification requirements, and the "arrested growth" philosophy that stands behind the motivation of the DAFs [Löb16, p. 26]. This leads to the following definition:

Definition 3.2 A *disbond arrest feature* is a design feature implemented in structural bonds for limiting the maximal damage size for any kind of growing damage in the bondline by ensuring damage in the bondline cannot pass the feature.

Overview of Disbond Arrest Features

Löbel [Löb16, pp. 24–25] and Sioutis and Tserpes [ST23] both motivate their reviews of DAFs with the certification requirements of the European and U.S. aviation authorities, illustrated in Sect. 2.1, and agree that DAFs are a very promising way to meet these. Their evaluation of existing concepts is thus considered relevant to the multifunctional bondline approach as well.

Löbel [Löb16, p. 31] classifies the disbond arrest concepts by two working principles. Extrinsic DAFs reinforce the bondline in an out-of-plane direction by providing a secondary load path. In contrast, intrinsic DAFs change the damage growth behaviour by altering the bondline itself via adhesive or surface modifications. Moreover, he states the following fundamental requirements for DAFs:

1. The feature shall be applied in a discrete manner. [...] The feature shall be combined with a state-of-the-art adhesive system.
2. The possible existence of weak bonds shall be taken into account. Eventual malfunction of the design feature itself shall thereby be avoided.
3. Both adherends and adhesive should not be weakened by the existence of the design feature. Consequently, the bond's overall strength shall not be reduced. If possible, the bond strength shall be improved by the presence of the feature.
4. The feature shall be applicable to all joint types (secondary bonding, co-bonding and co-curing).
5. The feature shall be producible with state-of-the-art manufacturing processes. [Löb16, p. 41]

Löbel [Löb16, pp. 32–34] reviews three extrinsic mechanisms. All of them are capable of arresting disbond growth, but they either disturb the composite joint similarly to mechanical fasteners or are not applicable to secondary bonding [Löb16, p. 40]. The latter is also true for the z-pinning approach, which Kadlec et al. [KRB20] use for their combination of a disbond arrest and an SHM feature [Löb16, p. 34].

Among the intrinsic mechanisms, Löbel [Löb16, pp. 34–39] evaluates adhesive toughening, surface modifications, self-healing, and multi-zone bondlines. Adhesive toughening reduces the stiffness and strength of the epoxy, and, at realistic loading conditions, cracks still grow [Löb16, p. 30]. Hence, Löbel [Löb16, p. 40] does not consider adhesive toughening as a DAF. The three reviewed surface modifications, on the other hand, show some improved properties, but Löbel excludes one of them because it is also not applicable to secondary bonding, and the other two because neither of them would account for the possible existence of weak bonds. Self-healing is evaluated as an alternative concept, which does not rely on arresting damages but repairing them automatically. It is excluded because the long reaction times necessary to complete the healing process would require awareness of the damage to ground the aircraft for that time. The reviewed mixed-zone adhesives include mixed-adhesive joints, selective toughening and filler elements. Mixed adhesives and selective toughening are both considered potential DAFs, but neither of them has been designed as one yet. In contrast, filler elements like peel stoppers are evaluated as a proven concept for a functioning DAF. However, Löbel [Löb16, p. 40] criticises that the filler elements would malfunction if they lie in a weak bond area themselves. He thus concludes that there is no intrinsic concept, which is actually designed as DAF and not prone to weak bonds. Consequently, Löbel et al. [Löb+16] have proposed the "Hybrid Bondline" as a new concept for DAFs.

Sioutis and Tserpes [ST23] review a slightly larger, yet similar, selection of DAFs as Löbel, but come to a different conclusion. In contrast to Löbel [Löb16], Sioutis and Tserpes [ST23] neither focus on the vulnerability of DAFs against weak bonds nor on their capability to overcome the disadvantages of mechanical fasteners for more efficient designs. Their major concern is that there is too little information on the effectiveness of DAFs under cyclic loading instead. As a result, they conclude that most of the approaches have the potential to be effective DAFs, but also point out the need for more tests under cyclic loading for a more profound understanding of their performance in aerospace applications [ST23, p. 15].

The fundamental requirements for DAFs defined by Löbel [Löb16, p. 41] (see above) are mainly in agreement with those of the multifunctional bondline defined in Sect. 3.1.4. Moreover, among the DAFs reviewed by Sioutis and Tserpes [ST23], there is no solution that promises a more functionally compliant design than Löbel's hybrid bondline. The hybrid bondline is thus analysed in detail in the following.

Hybrid Bondline

Figure 3.2 depicts a hybrid bondline concept in a bonded CFRP structure. The hybrid bondline consists of an epoxy adhesive and interim thermoplastic strips, subdividing the epoxy adhesive into multiple sections because the two materials are not mixable [Löb16, p. 42]. It is key for the concept that the polymer networks of different epoxy sections are completely disconnected for achieving an independent load path for each section [Löb16, p. 44]. Unlike the epoxy adhesive, the ductile thermoplastic shows low stiffness but a very high fracture toughness. Consequently, a crack cannot grow through it, and the initiation of a new crack in another epoxy section is required for further crack growth within the bondline [Löb16, p. 42]. A new crack initiation near the DAF after crack arrest is unlikely because the stress in the epoxy adhesive adjacent to the thermoplastic is reduced due to the thermoplastic lying at the new

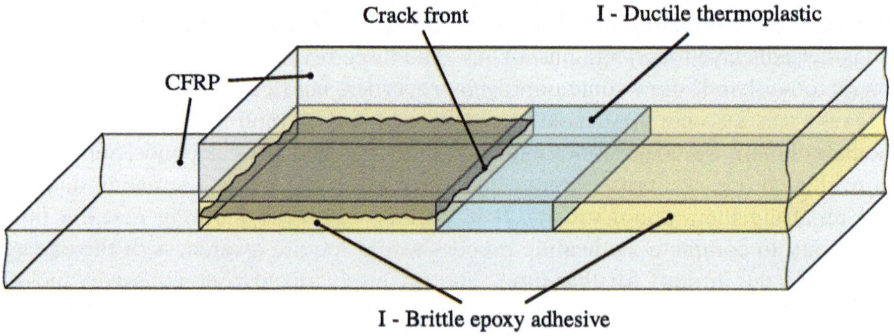

Fig. 3.2 Hybrid bondline concept, adapted from Löbel [Löb16, p. 42]

overlap edge. In fact, the stress in the epoxy adhesive adjacent to the thermoplastic is even significantly lower than the stress in the epoxy adhesive at the original overlap edge of the healthy bondline [Löb16, p. 166] because a mixed-adhesive joint is achieved after crack arrest [Löb16, p. 163].

Mixed-adhesive joints consist of flexible adhesives at the overlap edges and a much stiffer one in-between. As a result, the stress peaks at the overlap edges are significantly decreased, and a larger share of the load is carried by the adhesive in the middle. This leads to a more uniform load distribution within the bondline, and therefore an increased mechanical strength [Löb16, pp. 37–38]. The hybrid bondline is thus the implementation of the DAF based on the mixed-adhesive joints approach, which has not been accomplished before [Löb16, p. 38].

The thermoplastic material is not only supposed to split the epoxy adhesive into multiple regions but also to create a fusion bond[1] via the DAF. Only the combination of structural (adhesive) and fusion bonding forms the hybrid bondline. The main motivation behind this is avoiding weak bonds within the DAF. The fusion bond itself is less prone to weak bonds, as it does not rely on adhesion mechanisms [Löb16, p. 43]. It has to be pointed out, though, that an adhesive joint between the thermoset composite and the thermoplastic material is still formed in a co-curing process. A suitable melting temperature and the co-curing-capability to the thermoset matrix of the composite are thus two important additional requirements for the thermoplastic material [Löb16, pp. 48 and 52]. Löbel [Löb16, Chap. 8] observes strong adhesion of the thermoplastic PVDF material of the hybrid bondline to the matrix. According to Laurien et al. [Lau+18, p. 3989], the formation of semi-interpenetrating polymer networks during co-curing PVDF to an epoxy matrix causes the strong interfacial adhesion.

[1] Fusion bonding is also referred to as thermoplastic welding [Löb16, p. 6].

Crack growth tests over 10^6 cycles at maximum strain levels[2] in the composite adherends of 3000, 4000, and 5000 $\mu m\, m^{-1}$ under tension-tension load demonstrate the disbond arrest function of the hybrid bondline concept [Löb16, pp. 136–141]. Moreover, residual strength tests of samples with arrested crack show a higher strength than undamaged references, which proves the strength benefit after crack arrest due to the formation of a mixed-adhesive joint [Löb16, pp. 141–143].

Even though the functionality of the hybrid bondline has been proven, there are concept limitations [Löb16, p. 144]. Sub-surface cracks may occur, which cannot be arrested by the DAF. This means delamination failure has to be prevented by the joint design. Moreover, the hybrid bondline has not yet been studied under realistic environmental conditions for the application in composite aircraft.

Surface Toughening

The crack growth experiments of Löbel [Löb16, Figure 10.17] show that cracks in the hybrid bondline are already arrested at the slightly wider thermoplastic surface layer and not at the welded barrier. Schollerer et al. [Sch+20] takes on this effect, and investigates a variant of the hybrid bondline called surface toughening. The surface toughening principle relies on the thermoplastic surface layers co-cured to the composite alone. This means the extra thermoplastic layers forming the welded joint in the hybrid bondline are left out during the structural bonding process.

The main motivation behind using the surface toughening principle instead of the hybrid bondline approach is the much simpler and more robust manufacturing process [Sch+20]. Omitting the thermoplastic strips for the welded joint allows for using standard bonding processes. However, in contrast to the hybrid bondline, no welded joint is formed by this kind of DAF. This means the epoxy resin is not divided physically into multiple sections any more, and the risk of a weak bond within the DAF increases again.

The challenges posed by weak bonds and the robustness of the manufacturing process are both important. The question is if the more complex manufacturing process of the hybrid bondline for reducing the risks of weak bond formations at the DAF is worth the risk of deviating bondline thicknesses. Hart-Smith [Har21, p. 764] points out that the issue of weak bonds has to be addressed by a sophisticated bonding process, including a careful surface preparation, and that properly manufactured bonded structures are very durable. This remains true when manufacturing the hybrid bondline as well, since it cannot be expected that the welded joint of the hybrid bondline will provide a sufficient second load path if the entire bondline fails due to a weak bond. However, there is no experimental study showing how large a weak bond around the DAF of the hybrid bondline may become before it fails, too. It thus appears that focusing on a sophisticated bonding process with a well-prepared surface and a well-adjusted bondline thickness is the better approach for strong and durable bonds without weak bonds than a more complicated manufacturing process with uncertain bondline thickness, which reduces the risk of a weak bond in the area of the DAF only.

[2] Section 4.1.4 explains the choice of the maximum strain levels in detail.

Schollerer et al. [Sch+19] investigate the use of the surface toughening principle thoroughly at the overlap edge of bonded joints to improve the bond strength. Their experimental studies show an increase in strength of up to 80 % for SLS specimens with PVDF surface toughening at room temperature [Sch+19]. The first results of Schollerer et al. [Sch+20] also indicate that the surface toughening effect of PVDF is sufficient for arresting cracks at the limit load level in CLS specimens tested at room temperature. However, additional studies are required to confirm that and to evaluate the reliability of this method.

3.2.2 Review of Structural Health Monitoring of Structural Bonds

What are the approaches to SHM of structural bonds?

SHM systems are a promising addition to conventional NDI methods for composite aircraft. Lammering et al. [LDG18, p. 3] motivate the application of SHM by the need to detect damages in lightweight structures long before a structural failure occurs. In combination with CFRP structures, they consider SHM as a promising method, which increases the safety and reduces maintenance cost because the required inspection intervals for such structures could be reduced [LDG18, p. 4]. Baker and Scott [BS16, p. 544] agree with the point that SHM can be a less expensive alternative to conventional inspection methods. Moreover, they see the potential to apply SHM for inspecting parts that cannot be assessed affordably by conventional inspection methods, e.g. due to an excessive dis- and reassembly demand. Consequently, SHM can enable a damage-tolerance approach instead of a safe-life one for these parts by enabling NDI in the first place.

The benefits of SHM even go beyond online damage detection. Baker and Scott [BS16, pp. 561–563] discuss how SHM systems allow for operational monitoring as well. This includes operational loads, usage, and environmental monitoring. Data from the operational monitoring provides many new possibilities. For example, more realistic fatigue loads, usage spectra, and design limit loads for aircraft design and certification can be obtained, and the current structural integrity management of the entire fleet could also be validated and improved [BS16, pp. 561–562]. Baker and Scott [BS16, p. 586] thus conclude that SHM systems offer new possibilities for a cost-effective and efficient structural integrity management of composite aircraft due to their structurally integrated sensors.

Since joints are often the weak link in composite structures, SHM is of great interest for structural bonding as well [LP22, p. 2]. This section reviews SHM systems for structural bonds. Even though environmental monitoring seems attractive, too, the focus is on SHM for damage detection.

Guided Waves

The most investigated SHM method for composites in aerospace structures is guided waves, such as Lamb waves. It is attractive because it is capable of covering large areas and overcomes the through-thickness limitations of conventional ultrasonic testing [BS16, pp. 566–567]. The guided wave method relies on a sender and a receiver, which are often piezoelectric elements integrated into [MSW18, p. 463] or permanently attached to [SSW13, p. 105] the structure. Damage causes reflections, refractions, and mode conversions of the Lamb waves in elastic solids. These changes in the wave characteristics are taken for damage detection [LDG18, p. 6].

Weiland et al. [Wei+20] study a Lamb wave-based approach for SHM of a bonded CFRP structure representing the fuselage of an aircraft. The structure is a joint between a skin and a frame, with a network consisting of four piezoelectric sensors attached to the skin. Artificial damage is incorporated into the bondline by a release foil [Wei+20, Fig. 7]. Weiland et al. [Wei+20, pp. 2001–2002] use a modified Reconstruction Algorithm for Probabilistic Inspection of Damage (RAPID) of Yan et al. [YRR10] to detect the disbond in their study. This algorithm relies on a baseline or reference measurement to compare the captured signals to [YRR10, p. 378]. Weiland et al. [Wei+20, p. 2009] conclude that the modified RAPID algorithm is capable of detecting and localising the disbond, but for precise damage localisations, they suggest combining multiple damage indicators and see the need for optimising the sensor configuration and evaluation method. Weak bond detection is not considered in this article.

Ong et al. [Ong+18] propose a scattered wave method for detecting disbonds, but no weak bonds, in a lap joint. Their main goal is to demonstrate the feasibility of detecting disbonds with a guided wave approach that does not require a baseline reading. In their opinion, methods using baseline readings could give false alarms due to the ageing of the sensor or environmental conditions. Moreover, they demonstrate that the excitation frequency has to be selected carefully for the detection of adhesive disbonds.

There are also a few studies on weak bond detection with guided waves. Dugnani and Chang [DC17] propose a method for the in-situ detection of bondline degradation using piezoelectric transducers and the shear mode of guided waves. They present a two-dimensional analytical model to show how the absolute values of the admittance, the conductance, and the resonant frequency could be used as indicators for weak bond detection. Adams et al. [Ada+16] propose the application of a recursive feedback algorithm by Charutz et al. [Cha+13] to detect weak bonds and demonstrate the feasibility in a Finite Element (FE) analysis using a network of eight transducers. It is yet to be proven experimentally how reliable these methods are.

Strain Monitoring

Strain monitoring is another SHM approach, whose feasibility has been demonstrated on secondary bonds and composite bonded repairs [BS16, pp. 563–566]. Vanlanduit et al. [Van+21, p. 236] emphasise that measuring strains does not equal damage detection. This means that for strain-based SHM, it has to be assured that a change in the strain field at the sensor position correlates with the damage that shall be detected. This also implies strain changes due to damage are distinguishable from changes due to natural or environmental disturbances. In addition, Baker and Scott [BS16, p. 563] point out that the position of the initial stage of damage has to be known, to be able to place the sensors appropriately. They thus consider this method a "simple, robust approach in composite structures where the critical region is spatially well-defined and confined", and explicitly name bondline regions as an example [BS16, pp. 563–564].

Preisler et al. [Pre+15] look for an efficient method to assess damages in structural bonds by strain monitoring. The authors propose a smart SHM method, which means the strain and stress state of the structure is determined by a structural analysis in the sizing process. Critical regions are thus identified as well. They further suggest measuring the back face strain for single and double lap joints for damage detection. The superposition of tensile and compressive strains creates a point on the back face of the lap joints at which the net strain becomes zero. This point moves because a crack changes the effective overlap length, and thus, the load transfer. The strain at the original zero-strain position directly correlates with the damage size. Hence, Preisler et al. [Pre+15] claim a single strain gauge at this position suffices for damage monitoring. In addition, the authors discuss the application scenario of monitoring the entire critical region of a bonded repair patch of a damaged A350 door corner. Based on a two-dimensional numerical model, they propose an SHM concept requiring only two strain gauges (one at the top and the other one at the bottom of the surface). However, this method is only sensitive to damage close to the sensors. The authors, thus, extend the two-dimensional model to a three-dimensional one. This results in the authors' suggestion of placing one fibre optical sensor each at the bottom and the top of the repair patch for continuously monitoring the entire critical area. The major advantage of the smart SHM method in combination with the back face strain measurement seems to be the low number of strain sensors required for damage detection. However, the strain sensors cannot be used for simultaneous load monitoring of the bondline as the method relies on the sensors being completely unloaded as long as the bondline is intact.

Minakuchi et al. [MOT07] investigate a method to detect delamination of sandwich structures using FBG sensors. The sensors are embedded in the bondline of a honeycomb core and a face sheet. During the manufacturing process, the FBG sensors accumulate residual stress, which leads to a change in the reflected spectrum of the FBG in comparison to the one obtained before curing. Minakuchi et al. [MOT07] test their method in a quasi-static experiment under tensional loading. When delamination occurs, the residual stresses are released and, therefore, the reflected spectrum shifts back to the initial state before bonding of the FBG sensor. This recovery is obtained within a second after the delamination has occurred. Thus, Minakuchi et al. [MOT07] conclude that a delamination may be detected in real-time. However, it is not made clear if the structure around the sensor needs to be fully unloaded to be able to compare the spectra. Moreover, Minakuchi et al. [MOT07] report that the recovered spectrum does not match the initial one exactly. Consequently, the damage criterion is open to interpretation. To solve this, Minakuchi et al. [MOT07] propose using a small diameter chirped FBG sensor. Chirped FBG means that the distance between the Bragg grating elements grows monotonically, which leads to the reflection of a broad spectrum. After curing, the chirped FBG sensors show small peaks, especially at low wavelengths, which are attenuated by delamination. This additional information provides a more reliable damage criterion [MOT07]. Moreover, Minakuchi et al. [MOT07] observed that the spectrum of the FBGs changes in correspondence to the out-of-plane strains upon tensile loading. Hence, they claim this method can also be used for online monitoring of these strains. As crack propagation under cyclic loading is not investigated, it remains unclear how well this method works for detecting disbonds caused by fatigue, though.

Palaniappan et al. [Pal+08] and Guo et al. [Guo+09] use chirped FBG sensors for crack detection in structural bonds under fatigue instead. The measurement principle also relies on residual stresses, which have built up during curing and are released again upon failure of the bondline. Upon crack initiation, parts of the sensor are still fixed by the bondline and cannot release their residual stresses. As a result, the distances between the Bragg grating elements do not increase monotonically any more. Palaniappan et al. [Pal+08] use this effect, to detect the crack initiation in a single lap joint at either side of the sensor. In addition, Guo et al. [Guo+09] demonstrate the possibility to determine the crack length. Palaniappan et al. [Pal+08] and Guo et al. [Guo+09] both apply cyclic loads for damage growth in the bondline. It has to be pointed out, though, that the sensor data has been recorded at a constant load, and not during cyclic loading in both experiments. This raises the question if this principle is applicable for online monitoring as well, or only for measurements taken at standstill.

Yashiro et al. [YWS17] present a method for online monitoring of disbonds in structural bonds under cyclic loading. For that matter, three FBG sensors with a particularly long measurement grating of 25 mm are integrated into CFRP double lap joints. The sensors are embedded into the top layer of one adherend so that they are in direct contact with the bondline. The clue behind the measurement principle is the long measurement grating. Once the disbond starts growing, the sensor area above the disbond is unloaded as the adherend cannot carry the tensile load any more. Since the remaining sensor area still sees a strain, a second peak occurs in the spectrum of the FBG sensor. The ratio between both peak intensities correlates with the disbond length and thus provides a damage criterion, which is even resistant to partial failure of the optical fibre. Yashiro et al. [YWS17] derive the relation between the disbond length and the peak intensity ratio by an FE analysis for various disbond lengths. Based on these results, they fit the parameters of a polynomial function, which maps the disbond length to peak intensity ratio. In their FE analysis, they assume a rectangular disbond area. Yashiro et al. [YWS17] thus conclude that their method is specific to their specimens, but can be applied in general if similar step-like strain distributions occur.

Electromechanical Impedance

The electromechanical impedance method is another approach for SHM of structural bonds [TLd22, p. 229]. Measuring the frequency-dependent electromechanical impedance or admittance response of a piezoelectric transducer bonded to a structure reveals the resonance peaks of that structure. Hence, the measured frequency response of a damaged and a healthy state will yield different resonance peaks [RG17, p. 112]. An advantage of the electromechanical impedance method is that no external force excitation is necessary, as an in-plane strain excitation is used [Giu14, p. 512].

Zhuang et al. [Zhu+18] claim they can detect weak bonds in SLS specimens using the electromechanical impedance method. They place a 250 μm thick sensor in a bondline of an SLS specimen with aluminium adherends [Zhu+18, p. 1035]. For simulating weak bonds, different chemicals contaminate the interface before curing [Zhu+18, p. 1034]. The monitoring system is evaluated in static tests of weak and nominal bonds. For measuring the impedance, the tests are interrupted, and the specimens are unloaded [Zhu+18, p. 1035]. Zhuang et al. [Zhu+18, p. 1036] use a root-mean-square deviation method to create a damage index, which means the impedance from the measurements is compared to one of the healthy bonds. A change of the damage index by 2 % defines an arbitrary damage criterion. For most specimens, the damage index increases rapidly at $\approx$80 % of the normalised bondline strength. In addition to the experiments, Zhuang et al. [Zhu+18] provide an FE model of the bondline. The weak bonds are implemented via an additional layer between the adherend and the adhesive with a tuned stiffness, which equals 10 % to 90 % of the original stiffness [Zhu+18, p. 1038]. Zhuang et al. [Zhu+18, p. 1040] study their approach in fatigue tests and report that the damage index increases significantly at 88 % of the fatigue life in these. They complete their study by investigating the influence of the different sensor geometries and placements on the bondline strength and damage detection capability. They find that thinner sensors show an increase in sensitivity [Zhu+18, p. 1041], whereas there is no clear trend regarding the diameter of the sensors [Zhu+18, p. 1042]. The distance of the sensor to the damage does not yield a clear trend either, so that Zhuang et al. [Zhu+18, p. 1042] assume a complex interaction between the sensor and the structure that requires additional investigations. In that context, W. Roth and Giurgiutiu [RG17, p. 112] report that the detection capability of the electromechanical method is often limited to the local vicinity of the disbond. Zhuang et al. [Zhu+18, p. 1043] conclude from their results that their method is capable of detecting weak bonds. However, their experimental results indicate that the weak bond is not detected right away, but only after the load has reached $\approx$80 % of the strength of the bonded structure. A true weak bond detection, however, would require identifying it without loading the structure first. This does not seem feasible with an approach that uses the initial state of the bondline as a baseline measurement.

Malinowski et al. [MWO15] investigate whether weak bonds change the electromechanical impedance response in comparison to free sensors or sensors bonded to an undamaged and uncontaminated adhesive joint of composite structures. The weak bonds are created by contaminated interfaces, moisture and decreased curing temperatures [MWO15, p. 333]. Malinowski et al. [MWO15, p. 335] use the resonance related to the thickness mode and the frequency of the maximum peak as well as the root-mean-square value of the conductances as indices. The frequency index is incapable of separating all results, but it is sensitive to the level of weak bond [MWO15, p. 338]. The root-mean-square index, applied by calculating the relative value regarding the bonded sensors, also shows a change of at least 30 % if a weak bond is present. Hence, Malinowski et al. [MWO15, p. 343] conclude that a two-step method is feasible, and assume the implementation of this method in larger areas is possible as well.

Acoustic Emission

Acoustic emission is another SHM method that finds much attention for composite structures and structural bonds [LP22, p. 13]. The fundamental working principle is recording elastic waves, which are emitted by irreversible processes in the material, like crack nucleation and propagation, and evaluating the signal and corresponding waveforms [LP22, p. 175]. The sensitivity of acoustic emission is sufficient to detect micro-cracking, determining the damage mode, and three-dimensional localisation of the source [Agg+21, p. 176].

Teixeira de Freitas et al. [TZP18] investigate the possibility of using acoustic emission to distinguish between damage caused by weak bonds and nominal structural bonds. They investigate the acoustic emission method in DCB tests of structurally bonded CFRP specimens. Some samples contain stripwise local artificial weak bonds, created by contaminating parts of the surface with a release agent [TZP18, p. 7]. The weak bonds are clearly visible by the load drop in the load-displacement curve [TZP18, Fig. 13, p. 15], and the lower cumulative energy in the acoustic emission results [TZP18, p. 17]. This behaviour is explained by the different failure modes of the nominal and the weak bond. The former is cohesive failure, whereas the latter is adhesive failure [TZP18, Fig. 14, p. 16]. Teixeira de Freitas et al. [TZP18, p. 17] conclude that the different energy releases can be used to identify the failure mode. Reference measurements could therefore enable the identification and early detection of areas with poor adhesion. Moreover, they report that there is no acoustic emission activity during crack propagation through the contaminated strip of the DCB specimen. They conclude from this, that this information can be used for detecting weak bonds in-service where cracks may grow without acoustic emission activity [TZP18, pp. 18–19]. However, it is not discussed how a zero-activity with crack growth is distinguishable from a zero-activity due to no crack growth. Teixeira de Freitas et al. [TZP18, pp. 19–20] observe another effect in the wavelet transforms of acoustic emission signals. Only for the contaminated specimens is scattering observed at the beginning of the signals. In the four signal amplitude levels recorded, the scattering is the higher, the lower the signal amplitude, and at the highest amplitude it vanished completely. Teixeira de Freitas et al. [TZP18, p. 19] interpret that the weak bonds cause the scatter and that the high amplitude signals are not affected by the presence of weak bonds. In their conclusion, acoustic emission shows the potential for early detection of weak bonds, but further analyses are necessary for validating their preliminary results [TZP18, pp. 19–20].

Vanniamparambil et al. [Van+15] study an SHM system combining the acoustic emission and guided wave approaches in a bonded skin-stiffener structure made of CFRP. Two types of specimens are investigated under fatigue loading, one of which contains an artificial disbond. Three piezoelectric transducers are attached to the spar for the SHM systems [Van+15, pp. 290–291]. Vanniamparambil et al. [Van+15, p. 291] point out that both acoustic emission and guided waves in composites are very complex. Hence, an FE analysis for a more profound understanding of the wave propagation and optimisation of the damage detection is carried out as well. For the acoustic emission measurement, a pattern recognition method based on a novel frequency factor is proposed to enable automated damage detection [Van+15, p. 295]. For the guided wave analysis, a damage index relying on a baseline measurement is applied. However, this method is considered too complex to obtain a robust damage identification [Van+15, pp. 292 and 298–299].

3.2.3 Conclusions Regarding the Multifunctional Bondline

What are the shortcomings of the presented disbond arrest and SHM features, and how could combining both features overcome these?

The literature review illustrates different methods for the implementation of disbond arrest and SHM features in structural bonds. Nevertheless, there are certain shortcomings for either method. This section discusses these shortcomings and how the multifunctional bondline approach could overcome some of them.

With the hybrid bondline and surface toughening, there are two approaches to DAFs that are capable of arresting cracks without adding weight to the structure. Moreover, both processes are manufacturable using standard curing processes of composite joints. However, Löbel et al. [Löb+16] report a thickness mismatch between the epoxy adhesive (0.13 mm to 0.15 mm) and the DAF (0.25 mm) after curing, which they explain by possible squeeze-outs. Schollerer et al. [Sch+19] point out that these tolerance issues and undefined squeeze-outs make it difficult to manufacture hybrid bondlines according to the given certification standards in aerospace. Surface toughening does not suffer from the tolerance issues. It is thus expected to be more suitable for aerospace structures from a manufacturing perspective. However, it is not yet certain if the surface toughening effect alone is sufficient for a disbond arrest. The results of Löbel [Löb16, pp. 138–139] suggest that cracks are arrested at load levels relating to commonly used limit strains of 3000 μm but not for strains of 4000 μm or above. Clarifying the disbond arrest capability of the surface toughening effect thus requires further investigations.

The general concern of Sioutis and Tserpes [ST23] that there is too little information on the performance of DAFs under more realistic loading conditions seems justified. In this context, the risk of environmental degradation should be considered in investigations on the durability of structural bonds with DAF as well. These kinds of investigations are important for structural bonds with SHM features, too. In particular, for SHM systems based on reference measurements, the ageing of the sensors and their behaviour under changing environmental conditions has the potential to cause false alarms [Ong+18].

False alarms are only one of many challenges towards SHM in structural bonds of composite aircraft. Even though various methods can be considered suitable for damage detection in structural bonds on a laboratory scale, their application in practical structures reveals the complexity of damage detection in bonded composite structures. In this context, Li and Palardy [LP22, p. 20] conclude that single methods can distinguish good from weak bonds but not quantify the properties of the structural bond. Li and Palardy [LP22, pp. 20–21] thus suggest combining different methods to enable a more sophisticated quantification. Moreover, they propose the increased use of machine learning for damage evaluation with multiple methods, and an increased number of sensors at the bondline. However, obtaining a high damage detection probability without triggering false alarms is challenging in their opinion.

In addition to the challenges related to the SHM functions itself, the SHM system might also affect other parts of the bonded structure. For example, the wiring of the sensors is often accompanied by a significant weight penalty [MSW18, p. 471], which could jeopardise the gain in efficiency by joints without mechanical fasteners. Other than that, the effect of the integrated sensors on the bondline strength requires further investigation. For instance, the 250 μm thick piezoelectric transducers used by Zhuang et al. [Zhu+18] prevent thin bondlines. Considering that the integration of PVDF foils into the bondline can cause tolerance issues already (see above), it is very likely that even thin sensors will do the same.

The multifunctional bondline has the potential to solve some shortcomings of the disbond arrest and SHM features. In that context, a closer look at the certification requirements is helpful. The U.S. Department of Transportation Federal Aviation Administration [US 10, pp. 11–19] and European Union Aviation Safety Agency [Eur10, pp. 11–19] allow a damage tolerance approach for proof of structure. Figure 3.3 illustrates the relationship between three damage tolerance approaches, the damage size, the residual strength of the structure, and allowed inspection intervals for each approach. A safety feature, based solely on SHM, would make use of the slow growth approach. In that case, the SHM feature must ensure that damage is detected before the limit load capacity of the structure is lost [US 10, pp. 15–16], which implies all aforementioned difficulties regarding SHM of structural bonds. The DAF, on the other hand, would utilise the arrested growth approach. This requires damage to be fully arrested at a size where the limit load capacity of the structure is still guaranteed. In both cases, if damage that has reduced the residual strength below the ultimate load has been detected, the ultimate load capacity has to be re-established.

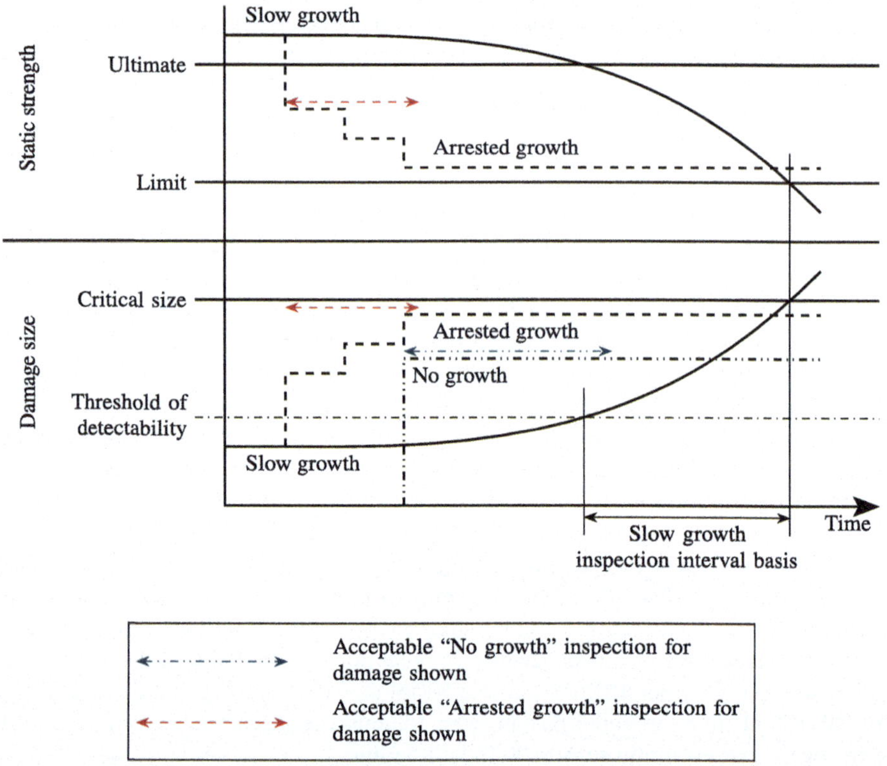

Fig. 3.3 "Illustrations of residual strength and damage size relationships for three different approaches to composite structural damage tolerance substantiation" [US 10, p. 16]. Reproduced figure

The combination of the disbond arrest and SHM features has the potential to lower the certification requirements for each feature. The risk that damage is not fully arrested but slowly grows further, could become acceptable if the SHM system identifies the event of the damage having reached the arrest features. This means the requirements for the disbond arrest function could be lowered so that it does not need to fully arrest damages that would cause the loss of limit load capacity, but to retard them sufficiently to make sure they are detected long enough before they reach that critical size. In turn, the challenge of SHM systems to characterise damage precisely to ensure a qualified estimation of the structural bond's residual strength could be overcome by the presence of the DAF. Dividing the bondline into sections allows for estimating the residual strength of the bondline by the number of intact sections, rather than complex estimations of the exact damage size and growth rate. Hence, the SHM system would only be required to identify damaged sections, which promises a much simpler evaluation. In terms of Fig. 3.3, the SHM system provides a shorter inspection interval than required for the acceptable arrested growth (dashed red arrow). This ensures damage is detected and fixed long before the critical size is reached. Hence, the scenario that damage has almost caused the structure to lose its limit load capacity would be prevented, thereby eliminating the requirement for the DAF to ensure damage close to the critical size is fully arrested. Investigating the feasibility of combining the disbond arrest and SHM functions in that manner is one core contribution of this thesis.

The lack of data about the durability under realistic loading and environmental conditions, however, is inherent to the multifunctional bondline as well. Hence, this will have to be solved by appropriate investigations.

The literature review, the three selected research questions, the meaning of functional compliance, and the research approach from Sect. 3.1.2 build the framework for the research hypotheses in this thesis. More specifically, a research hypothesis addresses each research question in the following sections. In turn, these research hypotheses lead to the fundamental working principles and guide the empirical investigations of the multifunctional bondline in Chap. 4.

3.3 Safe and Efficient Operation

Research Question 1

How to combine the disbond arrest, health monitoring, and structural bonding functions for a safe operation of bonded structures, while maintaining the mechanical strength and weight-saving potential of an ordinary, undamaged epoxy bondline?

Safe operation and low weight are two crucial technical requirements for a functionally compliant implementation of a multifunctional bondline. Only if the safe operation of the multifunctional bondline is provided at all times, certification for primary structures is achievable. Without weight benefits over established joining methods, the core motivation of achieving efficiency for the multifunctional bondline is not fulfilled any more. This emphasises the importance of finding an answer to Research Question 1 regarding a functionally compliant solution for the multifunctional bondline. This section thus proposes an answer to the first research question in the form of a research hypothesis.

3.3.1 Research Hypothesis 1: Functionally Compliant Operation by Multifunctional Disbond Arrest Features

Research Hypothesis 1: Functionally Compliant Operation
Embedding multifunctional disbond arrest features into epoxy bondlines at discrete positions enables the disbond arrest and SHM functions, while maintaining the mechanical strength and the weight-saving potential of an ordinary healthy epoxy bondline.

The literature review in Sect. 3.2 shows that the hybrid bondline and surface toughening approach are very promising to embed DAFs into epoxy bondlines without decreasing their mechanical strength. As they are both based on thermoplastics, they add much less weight to the structure than mechanical fasteners. In combination with the research approach from Sect. 3.1.4 to integrate the disbond arrest and health monitoring functions into an epoxy bondline, it seems a promising approach to integrate the sensors for an SHM system in the DAF as well. This way, the epoxy bondline remains mostly undisturbed, except for possible signal transmission paths to the sensor. As a result, it is expected to obtain a similar mechanical strength for the multifunctional and an undamaged pure epoxy bondline. The functional integration of the disbond arrest and SHM function leads to the novel MDAF.

3.3.2 Compliant Sensor Integration

Research Hypothesis 1.1: Compliant Sensor Integration
The integration of micro sensors into a DAF does neither harm the disbond arrest function nor decrease the mechanical strength of the structural bond. Moreover, the integrated micro sensors provide an unambiguous signal and do not fail under repeated mechanical loading of the structural bond.

Knowing the feasibility of integrating thermoplastic DAFs into epoxy bondlines without harming their initial strength directs the focus to the question of how to integrate a sensor for SHM purposes in a functionally compliant manner as well. The state of research in Sect. 3.2.2 provides many approaches to SHM of bondlines but only says little about the disturbance of sensors integrated into bondlines and vice versa. It is thus of interest what sensors are suitable to be integrated into a bondline or DAF.

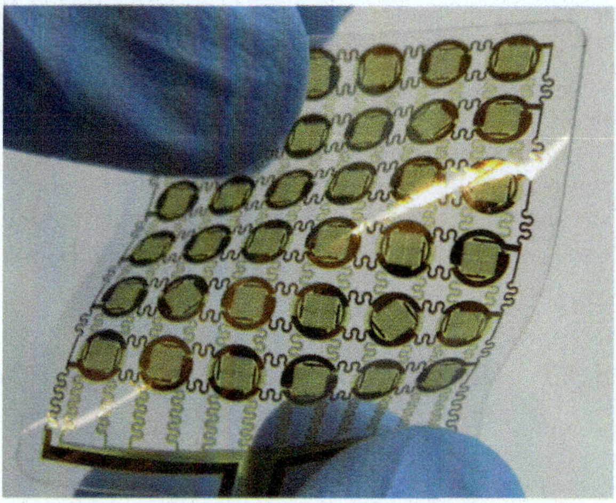

Fig. 3.4 Foil sensor array with polyimide substrate and strain gauges for bending measurements, similar to [KD16, p. 142]. Photograph provided by Eugen Koch

Li and Palardy [LP22, p. 2] name piezoelectric, fibre optical and nanocomposite sensors as commonly used ones for SHM applications in bondlines. Nanocomposite sensors are not considered in this thesis, as they are not a mature technology yet. Fibre optical sensors are often used for strain measurements in composites as well [BS16, p. 564]. Some of their advantages over electrical resistance strain gauges are their small size, which makes them easy to integrate, the possibility to integrate many discrete sensors on a single fibre, which leads to a lower weight penalty for the wiring, and a better durability and strain-to-failure [BS16, pp. 564–565]. Grundmann et al. [Gru+20] have found that FBG sensors in a polyimide coating and diameters of less than 80 μm neither significantly change the shear nor the fatigue strength upon direct integration into a structural bond. Investigating FBG sensors with outer diameters of 50 μm, 80 μm and 125 μm, they report an altered bondline thickness by all of them, but only for the 125 μm large sensor a weakened bondline in terms of decreased shear and fatigue strength is observed. As for piezoelectric transducers integrated into bondlines, Zhuang et al. [Zhu+18, p. 1039] report that it is still a main challenge to reduce their thickness, as the sensor they use is 250 μm thick. In that context, it is of high interest to create an intrinsic sensor by using the piezoelectric properties of the PVDF material, which forms the DAF of Löbel [Löb16]. This could minimise the disturbance by the sensor.

A promising alternative to the commonly used SHM sensors is foil sensors. These consist of a flexible thermoplastic substrate carrying micro sensor structures. As illustrated in Fig. 3.4, this technology enables very thin, flexible sensors [KD16]. Hence, foil sensors are attractive for reducing the disturbance of integrated sensors on the mechanical properties of a composite laminate by minimising the size of the disturbance, or wound, due to the integrated sensor [Lan+11, p. 4]. Kyriazis et al. [Kyr+21] show that replacing the commonly used polyimide substrate with PEI has great potential to significantly decrease the shear strength reduction caused by weak adhesion between the sensor grids and the matrix [Kyr+21, Fig. 8], and even to improve the peel strength compared to an undisturbed laminate [Kyr+21, Fig. 10]. They attribute this effect to the strong interface between the PEI substrate and the epoxy resin around the sensor grids. Hence, large proportions of the contact area between the sensor grids and the epoxy resin diminish the effect [Kyr+21, pp. 17–18].

The findings from the literature lead to the assumption that the integration of micro sensors into thermoplastic DAFs is possible without harming the disbond arrest or structural bonding functions. In question is the durability of the micro sensor structure itself, which is thus also included in Research Hypothesis 1.1.

3.3.3 Measuring Local Strain or Stress Gradients for Damage Detection

Research Hypothesis 1.2: Local Strain/Stress Gradient Measurements
Measuring the stress or strain gradient within the multifunctional disbond arrest feature only enables a clear judgement of the bondline status before damage reaches a critical size.

In addition to the approach to a functionally compliant sensor integration, it is important to define the SHM strategy. Following the method of Lang et al. [Lan+11] of a minimal invasive implementation by omitting everything not absolutely necessary for the sensor function, strain monitoring is very promising, as it provides a simple and robust method for bondlines [BS16, pp. 563–564]. Strain monitoring is thus used in this thesis for finding a functionally compliant solution.

Figure 3.5 shows the shear stress distribution of CLS specimens with different artificial crack lengths from the investigations of Löbel [Löb16, p. 162] regarding the hybrid bondline. (Similar CLS specimens are explained in Sect. 4.1.2.) In these investigations, crack lengths of 0 mm, 8 mm, 16 mm and 23 mm, measured from the overlap edge, are studied. The DAF is placed 25 mm away from the overlap edge. As already published in [Ste+21], a renewed FE evaluation based on the approach of Völkerink [Völ22] illustrates that not only the shear stress but also the longitudinal strain distribution (see Fig. 3.6) in the DAF changes significantly when the crack approaches it. This means that in the 23 mm crack result, there is a clear stress and strain gradient within the DAF. In contrast, in the no crack (0 mm) and 8 mm crack results there is no gradient because the stress and strain peaks occurring at the overlap edge or crack front decay to a plateau before the DAF region is reached. In the case of the 16 mm crack, the peaks have almost decayed upon reaching the DAF, but there is a slight gradient still. This leads to the assumption that measuring the strain or stress gradient within the MDAF only is sufficient for monitoring the bondline integrity by identifying damage which has grown near the MDAF.

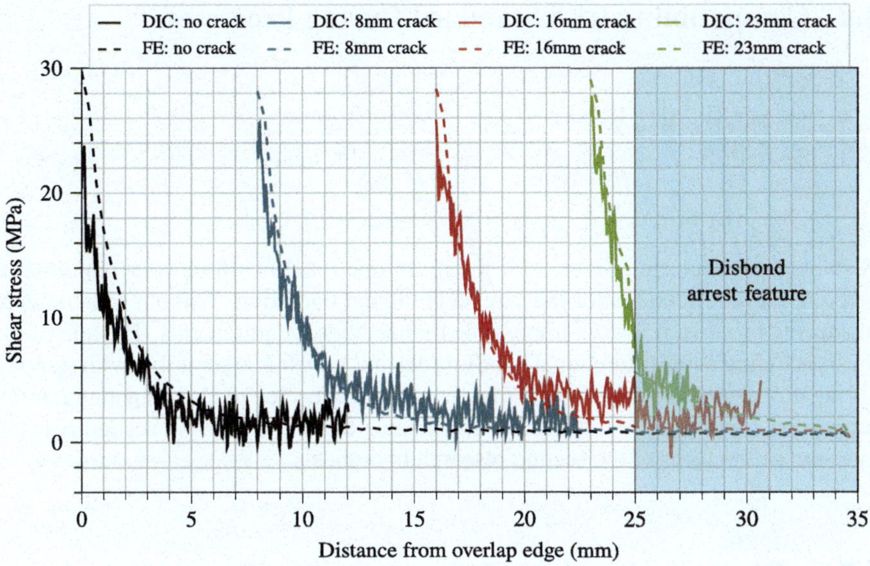

Fig. 3.5 DIC results obtained from Löbel [Löb16, p. 162] and FE results of the shear stress distribution within the hybrid bondline at four different crack lengths under 6 kN load. The DAF position starting at 25 mm is indicated by the blue background

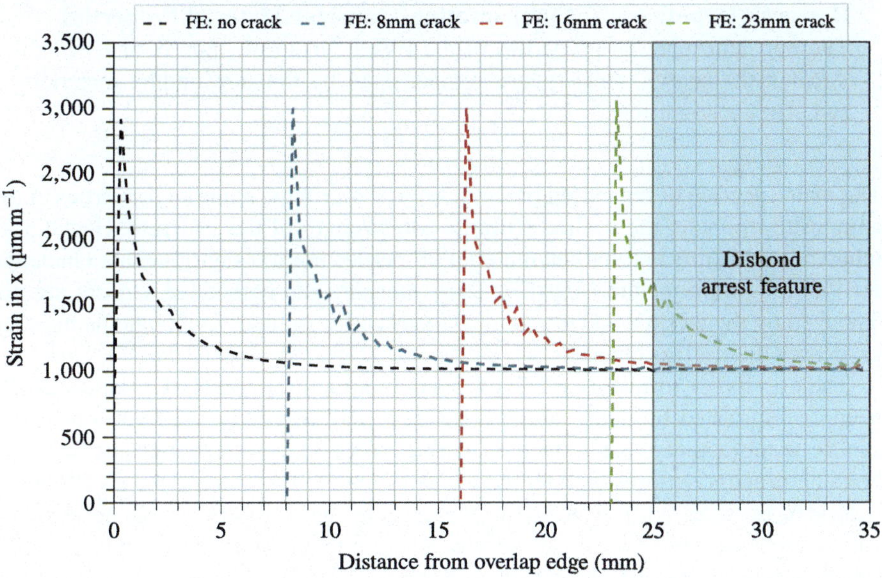

Fig. 3.6 FE results of the strain distribution within the hybrid bondline at four different crack lengths under 6 kN load. The DAF position starting at 25 mm is indicated by the blue background

3.4 Compatibility with Standard Curing Processes

Research Question 2
Is the multifunctional bondline manufacturable using standard curing processes?

A stable manufacturing process for the multifunctional bondline is as important as meeting all requirements for the safe and efficient operation. Without a stable manufacturing process, it is even impossible to meet these requirements. The aim of this thesis is to find a multifunctional bondline design, which is compatible with standard curing processes, to avoid new certification challenges due to altered manufacturing processes. Accordingly, this section formulates the answer to the second research question in the form of the second research hypothesis.

3.4.1 Research Hypothesis 2: Compensating Transient Material Properties with the Surrounding Materials

Research Hypothesis 2: Compensating Transient Material Properties
The transient material properties of the PVDF substrate upon curing, which threaten the sensor integrity, are compensated by the surrounding materials and process conditions.

The main function of the curing process is to transform the material properties of the composite to obtain the desired stiffness and strength of the structure. Hence, it is inherent to this process that the material properties are transient. A main challenge for the integration of the MDAF is thus to ensure the embedded sensors are not damaged by the thermal and mechanical loads acting on it, even though the material properties themselves change.

The possibility to integrate FBG sensors in structural bonds of composites has been demonstrated by many authors [LP22, p. 10]. Similarly, the hybrid bondline can be manufactured using standard curing processes, even though the adhesive bonding process requires high temperatures of 180 °C for welding the PVDF layers [Löb+16, pp. 231–232]. The research regarding the feasibility of using standard curing processes thus focuses on approaches to integrate foil sensors with a PVDF substrate or to utilise the piezoelectric effect of PVDF. This is motivated by the fact that the properties of the PVDF material are highly transient within the curing process because the material melts.

In terms of the foil sensors with PVDF subtrate, the transient stiffness threatens the support functions for the filigree sensor structures. For the intrinsic sensor, the transient piezoelectric effect threatens the sensor effect in addition to the risk of damaged sensor structures. The answer to Research Question 2 is, thus, that the manufacturability of the multifunctional bondline using standard curing methods demands compensating these effects by surrounding materials or the process conditions.

3.4.2 *Integrity of Integrated Sensors Throughout the Curing Process of Composite Structures*

Research Hypothesis 2.1: Sensor Integrity
The material surrounding the integrated sensors provides the support function for the filigree sensor grid to withstand the process-induced loads, at the time the PVDF substrate loses this functionality due to melting.

The main concern is the robustness of the filigree micro sensor structures of the foil sensors throughout the curing process, especially if the sensor substrate melts. It is highly questionable that the substrate is capable of fulfilling its support functions once it melted, or that the filigree sensor grid is mechanically robust enough to keep its shape under external loads. This means that once the PVDF substrate loses its support function because the stiffness decreases until it completely vanishes upon melting, this has to be provided by different means. Otherwise, the loads acting on the filigree sensors during curing might cause irreparable damage to them.

In the case of the multifunctional bondline, the caul plate could provide the support function. The caul plate is a metal plate placed on top of the laminate. It has the main functions of ensuring a smooth surface and pressing the MDAF into the CFRP material upon curing. As the MDAF is supposed to be placed in the interface of the CFRP and the epoxy adhesive, it is in direct contact with the caul plate. Moreover, the caul plate's stiffness is high throughout the entire curing process, in contrast to the stiffness of the PVDF and epoxy materials. Hence, it assumed that the caul plate can provide the necessary support function for the sensor structures upon melting of the PVDF. (Sect. 4.1.3 provides a detailed description of the manufacturing process.)

Another possibility is that the epoxy resin provides the support function when the PVDF melts. Löbel [Löb16, p. 100] reports how epoxy resin forms a sufficient number of cross-links upon reaching the melting temperature of PVDF to avoid mixing of the phases. This suggests the epoxy resin itself has built up a certain stiffness at the time the PVDF loses its stiffness due to melting. It thus appears feasible that the epoxy resin is capable of providing the support function for the sensor grids at the time the PVDF loses that capability.

3.4.3 Preserving the Piezoelectric Effect of PVDF

Research Hypothesis 2.2: Preserving the Piezoelectric Effect of PVDF
The hydrostatic autoclave pressure applied during the adhesive bonding process is sufficient to maintain the piezoelectric phases of polarised PVDF throughout the thermoset composite welding process.

The approach to utilise the piezoelectric effect of PVDF for the SHM system requires a method to create a piezoelectric phase of the PVDF material. According to Martins et al. [MLL14, p. 684], five different crystalline phases are known for PVDF, with the α-, β-, and γ-phase being the most commonly used ones. Solidification from the melt creates PVDF in the non-piezoelectric α-phase, which is normally transformed into the piezoelectric β-phase by a stretching process [Sin21, pp. 66–67]. According to Lovinger [Lov83, p. 1118], the mechanical elongation required for the phase transition is $\approx$300 %. Salimi and Yousefi [SY03, p. 703] even report stretching by 400 to 500 % at 90 °C are the optimal parameters for forming β-phase. However, the PVDF melts during the curing process, and stretching it after integration in the composite material is no longer possible[3]. Therefore, a different method is required to obtain a piezoelectric PVDF material.

Nakamura et al. [Nak+03] present a promising approach to maintain the β-phase of polarised PVDF throughout an annealing process. They report that annealing a stretched film at 182 °C while keeping its length constant forms pure crystalline β-phase. At slightly higher temperatures of 185 °C, the α-phase is formed instead [Nak+03, pp. 1703–1704]. As 180 °C is the maximum temperature in the curing and bonding process of the hybrid bondline [Löb+16, pp. 231–232], this method might be applicable for preserving the β-phase of polarised PVDF upon co-curing the PVDF to the composite adherends or curing of the epoxy adhesive as well. This requires the hydrostatic autoclave pressure to be capable of stabilising the PVDF during curing.

3.5 System Complexity

Research Question 3
How to manage the system complexity of the multifunctional bondline?

[3] Strain-to-failure of unidirectional carbon fibre composites is in the order of 1 % [Bar18, p. 32]

As discussed before, the system complexity is important in the scope of functionally compliant multifunctional bondlines. This is not only evident from the need to solve possible conflicts between efficient and safe designs as discussed in Sect. 3.1.3, but also from the findings in Sect. 2.1 that the complexity of structural bonds is one reason that prevents its broader application. Therefore, this section presents the third research hypothesis as an answer to the third research question.

3.5.1 Research Hypothesis 3: Complementing Functions Enable Lowering the Individual Requirements

Research Hypothesis 3: Complementing Functions
The complementing characteristics of the disbond arrest and SHM functions enable lowering the requirements for each function. In turn, this allows for a reduced complexity of the corresponding features.

The main goal of the multifunctional bondline is to enable the safe operation of structural bonds by the combination of the disbond arrest and SHM functions. As discussed in Sect. 3.2.3, the fundamental idea is combining two separate safety functions, which have the potential to compensate for each other's shortcomings. This means that rather than fulfilling the certification requirements for either a disbond arrest or an SHM feature completely, both requirements are only fulfilled partially. The overall safety requirement is, thus, only met because both safety functions are present and complement each other. In other words, the requirements of both individual functions are reduced in comparison to a safety mechanism that relies solely on one of these functions. Hence, a functionally compliant design might be found more easily in comparison to a bondline with only one of these safety mechanisms. This assumption implies that it is easier to harmonise the disbond arrest, SHM, and structural bonding functions at reduced requirements than harmonising either the disbond arrest or the SHM function with the structural bonding one at increased requirements for the safety functions.

The idea of reducing the requirements for the individual safety functions also has the potential to reduce the system complexity. If the requirements for the individual functions are lower, simpler solutions are likely sufficient to meet them. Consequently, the corresponding features could be designed less complex. The possibility of managing the system complexity through reduced requirements of complementing functions is investigated based on the disbond arrest and SHM functions in this thesis.

Since disbond arrest and SHM are safety features, the idea of being able to reduce the requirements for each function assumes that both safety features function in parallel. Only then, the shortcomings of one safety feature can be compensated by the other function. This also implies that a failure of one safety feature does not cause an immediate failure of the other feature.

3.5.2 *Disbond Arrest by Surface Toughening*

Research Hypothesis 3.1: Disbond Arrest by Surface Toughening
The surface toughening effect is sufficient to arrest disbonds within a structural bond for detecting and reacting to damage before it reaches a critical size.

The surface toughening effect for disbond arrest is a good example of how the combination with SHM could lead to a better and less complex multifunctional bondline. The literature review in Sect. 3.2 illustrates that surface toughening promises a simple and robust manufacturing process. One advantage in comparison to the hybrid bondline is that the bondline thickness is not influenced, avoiding tolerance issues upon manufacturing (see Sect. 3.2.3). As a result, the strength of the bondline is less likely reduced by a deviation in the bondline thickness if surface toughening is applied as DAF. However, it is uncertain if the surface toughening effect alone is sufficient for damage arrest under fatigue loading. It is thus questionable if the surface toughening effect alone is capable of meeting the certification requirements. The claim of this thesis is that due to the SHM feature, the disbond arrest function does not have to ensure damage is fully arrested, as long as its growth is hindered enough to detect and react to the damage before it reaches a critical size. Hence, the requirements for the DAF are lowered, which makes it more likely that the benefits of the surface toughening approach regarding the less complex manufacturing process can be utilised while ensuring a safe operation.

3.5.3 *Monitoring the Integrity of Bondline Sections Instead of Damage Initiation and Size*

Research Hypothesis 3.2: Monitoring Bondline Sections
The combination with disbond arrest features enables SHM systems which only require detecting damaged bondline sections instead of damage initiation and the exact damage size.

The potential to lower the requirements for the individual functions by combining disbond arrest and SHM features applies to the SHM function as well. As concluded from the literature review in Sect. 3.2.3, the requirement for a very precise damage detection and characterisation could be reduced to identify damaged bondline sections only due to the presence of the DAF. This holds great potential to decrease the complexity of the SHM feature, not only in terms of the damage evaluation method but also the number of required sensors.

3.6 Working Principles of Bondlines with Multifunctional Disbond Arrest Feature

What are the working principles of the multifunctional bondline approach utilising MDAFs?

The research hypotheses lead to the three working principles of the multifunctional bondline approach of this thesis:

1. Joining the adherends by structural adhesives
2. Preventing damage from spreading rapidly due to DAFs formed by either surface toughening or the hybrid bondline approach
3. Identifying damaged bondline sections through shear stress or strain gradient measurements with sensors integrated into these DAFs

The working principles incorporate the idea of implementing a novel MDAF for obtaining a functionally compliant multifunctional bondline. This section introduces two basic concepts for MDAFs. These serve as a starting point for investigating the functional compliance of this approach by testing the research hypotheses experimentally. Figure 3.7 illustrates both concepts.

The first concept consists of surface toughening elements for the disbond arrest and embedded micro sensors for the SHM function, as illustrated in Fig. 3.7a. It is based on the idea from Research Hypothesis 1.1 to use FBG or foil sensors for the SHM feature to keep the disturbance of the bondline and DAF due to the sensor at a minimum. Hence, the PVDF either functions as a substrate for the foil sensors or has no function for the sensor system in the case of the FGB sensors. In both cases, the phase of the PVDF material is not of importance. As shown, the PVDF strips and micro sensors of the MDAF are pressed into the CFRP adherends. As a result, the thickness of the epoxy bondline should not be influenced.

The second concept utilises the hybrid bondline for disbond arrest and an intrinsic sensor, using the piezoelectric properties of PVDF, for SHM. As depicted in Fig. 3.7b, a continuous PVDF barrier is extended by two electrodes for the implementation of the sensor system. The PVDF in-between the electrodes is the sensor material. As a result, except for the electrodes and signal transmission paths, no extra material is necessary. Hence, the disturbance of the DAF due to the sensor is also expected to be

(a)

(b)

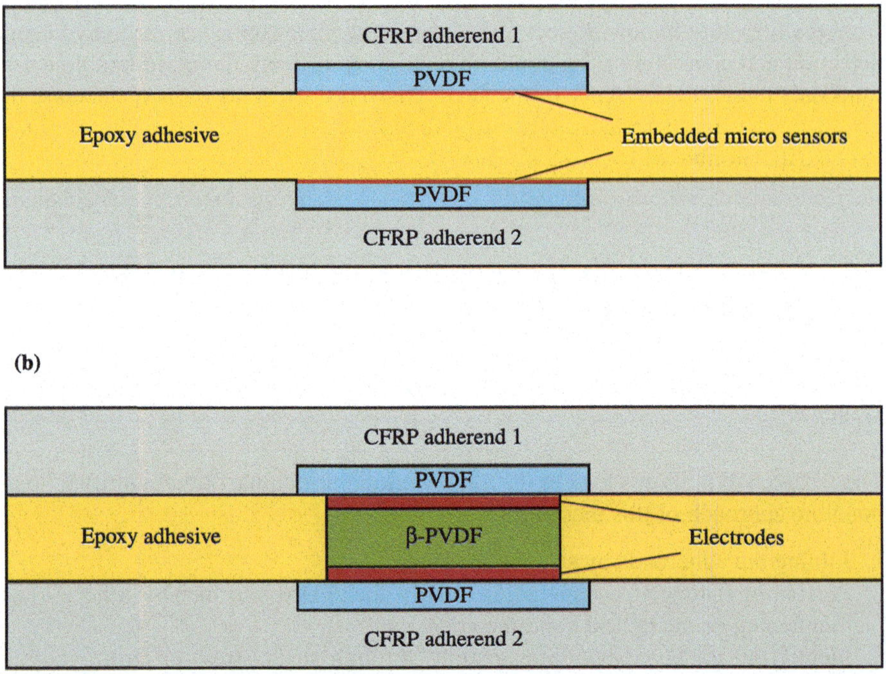

Fig. 3.7 Basic MDAF concepts **a** PVDF surface toughening strips for disbond arrest and embedded micro sensors for SHM **b** PVDF-based hybrid bondline for disbond arrest with intrinsic piezoelectric PVDF sensor, including electrodes, for SHM

minimal as long as the electrodes still allow for a thermoplastic weld of the PVDF layers.

For this concept, the phase of the PVDF material is of great importance because the β- and the γ-phases show a piezoelectric effect only. Figure 3.7b illustrates the piezoelectric phase as β because this is the preferred one due to its larger piezoelectric effect (see Sect. 3.4.3). Accordingly, Research Hypothesis 2.2, claiming that the piezoelectric effect of PVDF can be maintained throughout the manufacturing process of the multifunctional bondline, is strongly connected to this concept.

Both concepts illustrate the fundamental idea from Research Hypothesis 1 to keep the disturbance of the epoxy adhesive at a minimum. Due to the high level of functional integration, only little material is added to the bondline, and by placing the multifunctional bondlines away from the stress concentration at the initial overlap edge, the load transfer of the undamaged bond is expected to remain unchanged.

The open questions regarding the functional compliance of the multifunctional bondline approach of this thesis come into view through these concepts as well. Sorted by the three research hypotheses and expanded by some additional design questions, these are the following:

1. Safe and Efficient Operation
   - Does the multifunctional bondline provide limit load capacity repeatedly and ultimate load capacity?
   - Are the MDAFs capable of arresting damage?
   - Is the principle of measuring the stress/strain gradient sufficient for identifying damaged bondline sections?
   - Do the sensors withstand repeated mechanical loading?
2. Standard Curing Processes
   - Are the materials surrounding the embedded sensors or electrodes capable of supporting the sensor grids sufficiently to ensure their integrity throughout the curing processes?
   - Is the autoclave pressure sufficient to maintain the piezoelectric phase of the PVDF throughout the curing process for the intrinsic sensor concept?
3. Complexity Management
   - Are damaged bondline sections detected before damage spreads throughout multiple sections, or is the damage arrested long enough for the sensor to detect it? This implies the question if surface toughening without embedded sensors provides a disbond arrest function as well.
   - Is the detection of damaged bondline sections in combination with the DAF sufficient for ensuring the integrity of the bondline?
4. Multifunctional Bondline Design
   - What is a suitable design for the MDAF that enables measuring the stress/strain gradients?
   - How to contact the sensors of the MDAF?

3.7 Summary

Functional compliance of the multifunctional bondline regarding the long-term objective of improving the sustainability of aircraft can only be verified if all requirements from all life cycle phases of the aircraft are considered. As with any design task, this is a highly iterative process, which requires a more profound understanding of the basic working principles of the multifunctional bondline first. Providing the latter is the focus of this thesis. For that purpose, it is appropriate to define a more specific meaning of functional compliance for investigating the feasibility of combining the functions of the multifunctional bondline by introducing the system boundaries of an overlap joint.

Within these system boundaries, functional compliance means that the structural bonding, disbond arrest and SHM functions need to be harmonised in terms of providing a safe and efficient operation of the aircraft, and that the multifunctional bondline shall be manufacturable using standard curing processes. In addition, the complexity of the system shall be managed, not only to fulfil all functions in the first place but also to ensure the efficiency and safety of the multifunctional bondline. This is important because a higher system complexity might increase the efficiency and the risk of unexpected failures of a system at the same time. Accordingly, three fundamental research questions are formulated for studying the feasibility of multifunctional bondlines to provide a safe and efficient joining technology for composite aircraft.

Three main research hypotheses propose an answer to these questions. In summary, these claim that the functionally compliant operation of the multifunctional bondline is possible by the integration of a novel MDAF, that the multifunctional bondline is manufacturable using standard curing methods, and that the comprehending characteristics of the disbond arrest and SHM functions allow reducing the requirements, and therefore complexity, of the corresponding features. The research hypotheses lead to the working principles of the multifunctional bondline approach of this thesis. These are joining the adherends with state-of-the-art structural adhesives, preventing damage from spreading rapidly due to DAFs formed by either the hybrid bondline or surface toughening, and identifying damaged bondline sections by shear stress or strain gradient measurements with sensors integrated into these DAFs. Regarding the complexity reduction, these working principles incorporate the idea of using the complementing characteristics of the disbond arrest and SHM functions to reduce the requirement for both corresponding features. Moreover, the surface toughening approach is evaluated as the more promising one in terms of a less complex and more stable manufacturing process.

Based on the research hypotheses and working principles, two basic concepts for the MDAF are derived. The first one consists of surface toughening elements with embedded micro sensors. The second one proposes to add electrodes to a PVDF-based hybrid bondline to use the piezoelectric effect of this material for an intrinsic sensor.

Chapter 4
Functional Compliance of Bondlines with Multifunctional Disbond Arrest Features

Is the proposed multifunctional bondline approach functionally compliant regarding the system boundaries of an overlap joint with composite adherends?

This chapter deals with the experimental investigations for testing the research hypotheses. It starts with illustrating the specimen selection and investigation methods. This includes the design approach for the MDAF. Thereafter, a section for each of the three main research hypotheses presents and discusses the corresponding results. The chapter closes with evaluating the functional compliance of the MDAFs based on the experimental investigations. Parts of this chapter have been published previously in [Hei+21], [Hei+22], and [Vas+21].

4.1 Specimen Selection and Investigation Methods

What methods are used to study the proposed multifunctional bondline approach?

The starting point for the specimen selection and investigation methods are working hypotheses, defining the core investigation methods. Based on these, this section introduces the specimen design and a detailed description of each investigation method.

J. Steinmetz, *Functional Compliance of Multifunctional Bondlines with Disbond Arrest and Health Monitoring Features for Safe and Efficient Composite Joints*, Mechanics and Adaptronics, https://doi.org/10.1007/978-3-031-72422-0_4

4.1.1 Working Hypotheses

What are the key assumptions for investigating the functional compliance of the proposed multifunctional bondline approach?

For investigating Research Hypothesis 1 regarding a safe and efficient operation, a setup is desirable for testing the structural bonding, disbond arrest, and SHM functions simultaneously under realistic loading conditions. For a similar motivation, Löbel [Löb16, pp. 125–126] suggests fatigue tests using CLS specimens when studying the disbond arrest capability of the hybrid bondline. He points out three advantages: representing joints in doubler-to-skin configurations, the possibility to apply realistic load magnitudes, and an unambiguous crack growth direction. Hence, crack growth tests with CLS specimens are also considered very valuable for studying the functional compliance of the multifunctional bondline.

Regarding the open questions for the safe and efficient operation, the crack growth tests with CLS specimens enable evaluating the load capacity, disbond arrest capability, stress/strain gradient measurement principle, and sensor integrity for different multifunctional bondline designs. For the evaluation of the measurement principle, the crack length has to be known and correlated to the sensor outputs. Optical crack length measurements on one side of the specimen provide this information. As the crack length inside the specimen cannot be guaranteed to be the same across the entire bondline depth, extra quasi-static tests of CLS specimens with different artificial crack lengths shall support studying the correlation between the crack front and the sensor outputs. Moreover, quasi-static residual strength tests after the crack growth test shall allow for assessing the influence of the (M)DAFs on the static mechanical strength of the bonded specimen.

The CLS crack growth tests also enable investigating Research Hypothesis 3 about the reduction of the requirements for the disbond arrest and SHM features. Using specimens with surface-toughening-based MDAFs answers the open question of whether damages are detectable before they pass through the feature. In that context, the optical crack length measurements are an important tool, as they not only provide the information at what crack length the damage becomes visible but also if and how long it is arrested. Moreover, DIC stress/strain measurements enable a gradient measurement across the entire length of the specimen, and outside the bondline. A comparison of the DIC strain measurement and the discrete one from the sensors inside the MDAF demonstrates if there are cases when other sensor positions would have been necessary for proper damage detection. This answers the open questions if the detection of damaged bondline sections in combination with the DAF is sufficient for ensuring the bondline integrity in general, and if the proposed disbond arrest and SHM features detect damage before it spreads across multiple sections.

The implementation of the MDAFs in the CLS specimens requires answering the open questions regarding a suitable design for measuring the stress/strain gradients and contacting the sensors. For the stress/strain gradient measurement, one option is distributed micro strain gauges. Regarding the MDAF concept with embedded micro sensors (Fig. 3.7a), this could either be FBG sensors with multiple measurement gratings or multiple electrical resistance strain gauges. As for the concept with a piezoelectric sensor (Fig. 3.7b), an interdigital electrode could enable gradient measurements. For the connection of the sensors, the basic assumption is that a thin conductor path alongside the bondline enables a stable signal transmission to the outside of the bondline, i.e. in front of the overlap edge, and that this causes a negligible disturbance to the bondline.

The approach for testing Research Hypothesis 2.1 is twofold. The first investigation is melting foil sensors on a PVDF substrate under a microscope. Placing a cover glass on the sensor grid of some specimens allows for simulating a stiff material surrounding the sensor structures in the curing process. This way, it becomes visible whether the melting PVDF causes a load on the sensor grid, and if the surrounding material is capable of supporting the sensor grids sufficiently during the process. In the second step, the manufacturing process of the CLS specimens is used for studying the sensor integrity of the embedded sensors throughout the curing process. Function tests of the embedded sensors before and after each manufacturing step provide information on the feasibility of using standard curing processes for the multifunctional bondline. In the case of the foil sensors, the resistance of the micro strain gauges provides information on the sensor integrity, for example. This answers the open question of whether the embedded micro sensors stay intact throughout the actual curing processes.

Testing Research Hypothesis 2.2 and answering the open question regarding the possibility of preserving the piezoelectric effect throughout the curing processes requires a different method. Fourier-transformed infrared spectroscopy (FTIR), X-ray diffraction (XRD), and differential scanning calorimetry (DSC) are common techniques for identifying the phases of PVDF [MLL14, pp. 684–688]. However, none of these is applicable after the integration of the intrinsic sensor into the bondline. Hence, the proposed method in this thesis is applying the curing processes to a polarised PVDF foil rather than a complete specimen, and measuring the phase contents before and after. This way, the PVDF is constrained by the tooling and vacuum setup only instead of the surrounding composite. It has to be assumed at this point that this does not make a significant difference in terms of the phase changes. Vasic [Vas21] has developed and carried out the according investigation in the scope of his Master's thesis. These investigations are also published in [Vas+21].

All approaches described in this section are summarised in the form of working hypotheses in the following:

Working Hypothesis 1: Crack Growth Tests with CLS Specimens

Crack growth tests of CLS specimens with MDAFs are suitable to test all functions for the functionally compliant operation of the multifunctional bondline simultaneously. This test is therefore ideally suited for investigating Research Hypothesis 1 about achieving functional compliance during operation by the implementation of MDAFs.

Working Hypothesis 2: Optical Crack Growth and Strain Measurements

Optical crack growth and the DIC stress/strain measurements allow for the evaluation of the MDAFs. Foremost, it enables correlating the sensor data with the crack length during the crack growth experiments. Therefore, the capability of the sensor system to detect damage near the MDAF can be tested. Repeated crack length measurements throughout the test show at what size damage is detected and how well it is arrested. The DIC measurements enable checking the measurement principle and allow stress/strain measurements at arbitrary positions. Bringing this information together also enables discussing Research Hypothesis 3 regarding the possibility of reducing the system complexity due to the complementing nature of the disbond arrest and SHM functions.

Working Hypothesis 3: Quasi-Static Tests of CLS Specimens

Quasi-static testing of CLS specimens with defined artificial disbonds allows for a more precise evaluation of the stress/strain gradient measurement principle. Moreover, quasi-static residual strength tests after the crack growth tests provide insight into the influence of the different (M)DAFs on the static strength of the bonded composite.

Working Hypothesis 4: Distributed Micro Strain Gauges

Measuring the stress or strain gradient within the MDAF is possible by distributed micro strain gauges.

Working Hypothesis 5: Sensor Connection Alongside the Bondline

The sensors of the MDAF can be connected via thin conductor paths alongside the bondline. This way, the signal can be let across the overlap edge to the open surface of one adherend where it can be further contacted. This allows for a stable connection at a negligible disturbance of the bondline.

Working Hypothesis 6: Melting of Foil Sensors under a Microscope

Melting PVDF-based foil sensors under a microscope with different boundary conditions allows for studying Research Hypothesis 2.1. Placing a cover glass on top of the foil sensor simulates the surrounding structure, e.g. the caul plate. Comparing the behaviour of covered and uncovered foil sensors illustrates the feasibility of providing the support function of the sensor structure upon melting of the PVDF by the surrounding material.

Working Hypothesis 7: Manufacturing CLS Specimens

The compatibility of the multifunctional bondline approach with standard curing processes can be tested through the manufacturing of the CLS specimens with different MDAF types. Checking the sensor integrity after each manufacturing step enables testing Research Hypothesis 2.1 regarding the feasibility of embedding the SHM system into a DAF using standard curing processes.

Working Hypothesis 8: Curing Processes with Piezoelectric PVDF

The feasibility of utilising the piezoelectric effect of the PVDF material for SHM can be investigated by subjecting polarised PVDF foils to different curing processes. Analysing the phase content of the PVDF before and after shows the validity of Research Hypothesis 2.2 to be able to preserve the piezoelectric phase by the hydrostatic autoclave pressure.

4.1.2 Multifunctional Disbond Arrest Feature and Cracked Lap Shear Specimen Design

What do the specimens for investigating the functional compliance of the MDAFs look like?

The experimental investigations of the multifunctional bondline require a design of the CLS specimen with integrated MDAF. This section presents the designs of the MDAFs and CLS specimens used for testing the research hypotheses.

In the scope of this thesis, four MDAF designs are investigated. Figure 4.1 illustrates CLS specimens with these designs. The side view of the specimen depicted in Fig. 4.1a is the same for all designs. The different MDAFs are depicted in parts b to d. The first design consists of surface toughening elements for disbond arrest, on which an FBG array is placed (Fig. 4.1b). Here, the sensors and DAFs are integrated into the bondline by two separate elements. The other three designs are different generations of a smart inlay made of micro strain gauges on a thermoplastic substrate. The smart inlay is a foil sensor that functionally integrates the sensor and disbond arrest functions. The main differences between the smart inlay generations are the positioning of the conductor path (Figs. 4.1c and d), the thermoplastic foil used as substrate, and the presence of a second foil as cover for the sensor structures. Before a detailed explanation of these differences, the layout of the CLS specimen and general aspects of the MDAFs are presented.

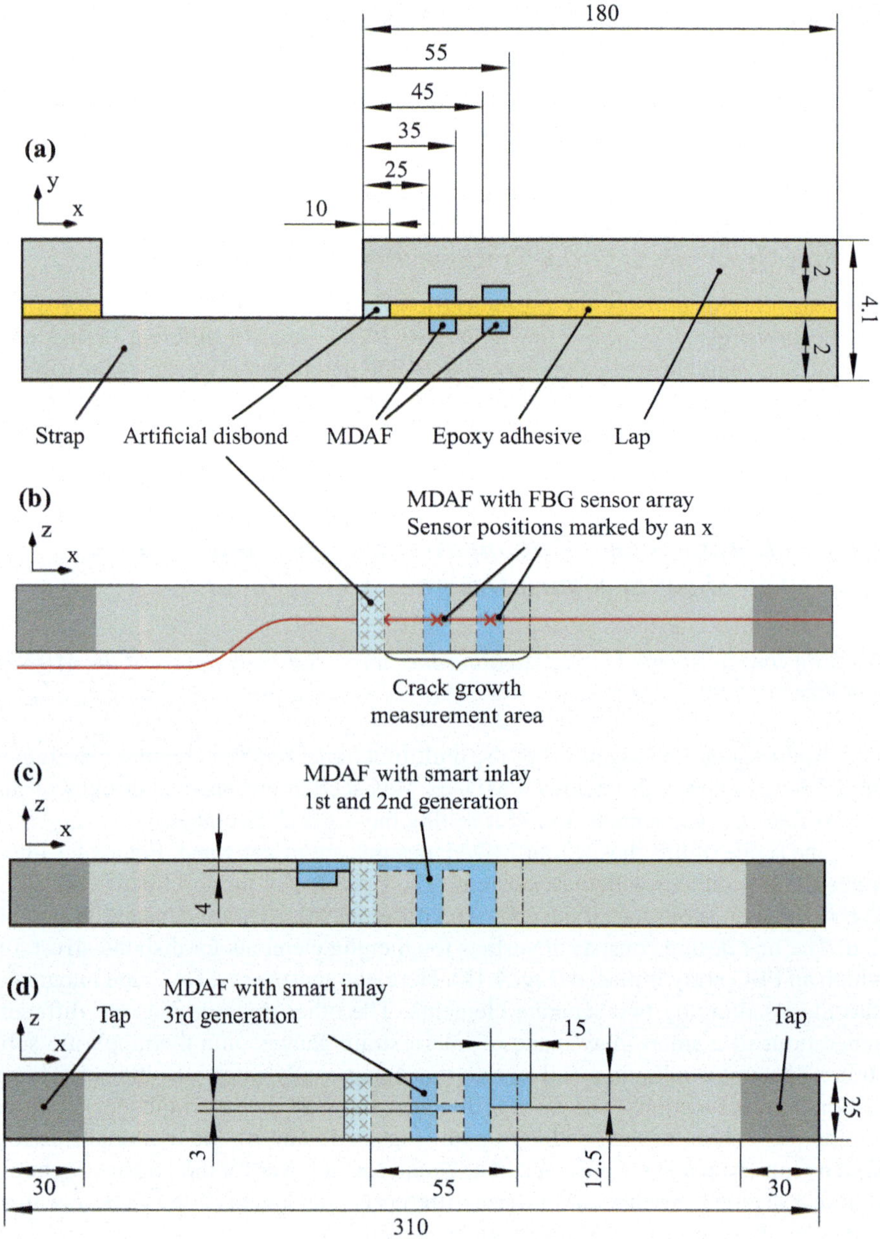

Fig. 4.1 CLS specimen design with four different MDAFs **a** Side view of all MDAFs **b** Top view of the FBG MDAF. The FBG fibre is in the interface of the MDAF and the epoxy adhesive **c** Top view of the first- and second-generation smart inlay MDAFs. The detailed sensor layouts and stacking sequences in y-direction are shown in Figs. 4.3, 4.4, and 4.5 **d** Top view of the third-generation smart inlay MDAF

General Cracked Lap Shear Specimen Design

The adherends of the CLS specimen (strap and lap) are made of sixteen plies unidirectional prepreg[1] with $[0/+45/90/-45/0/+45/90/-45]_s$ stacking sequence. The choice of 0° layers at the surfaces of the adherends shall avoid delamination failures close to the bondline. The strength and damage mode of structurally bonded composites under fatigue loading depends on the orientation of the plies at the interface [BS16, p. 441]. While a 0° ply shows only damage in the adhesive, delaminations in the laminate occur for ±45° and 90° configurations in addition to the damage growth in the adhesive. In the 90° case, the damage mode even changes to pure delamination after some distance.

The adhesive is an epoxy film adhesive[2]. The usage of film adhesives promises a better reproducibility of the bondline in terms of thickness in comparison to paste adhesives [Löb16, p. 9]. Moreover, an artificial disbond is created by placing a release foil in-between the adhesive and the strap at the first 10 mm of the overlap edge. This ensures an unambiguous crack initiation.

Two-Stage Multifunctional Disbond Arrest Features Design

All MDAFs consist of a two-stage design for damage detection. As shown in Fig. 4.1, the two-stage MDAFs are made of two 10 mm wide DAFs with integrated sensors in the strap. In addition, there are two DAFs without integrated sensors in the lap. The DAFs are surface toughening elements co-cured to the adherends (see Sect. 4.1.3) and, thus, lie directly in the interface between the adherends and epoxy adhesive. The distance between the DAFs in each adherend is another 10 mm.

The idea behind the two-stage design is to create a simple and robust measurement principle for the strain gradient, which relies on the minimal overlap length for load transfer in a doubler configuration. According to a load transfer model for elastic adhesives [BS16, p. 411], this load transfer length

$$L_{\min} = \frac{3}{\beta} \tag{4.1}$$

is the length required to transfer 95 % of the load from the strap to the lap in a CLS specimen. It depends on the estimate

$$\beta^2 = \frac{G_\mathrm{A}}{t_\mathrm{A}} \left[\frac{1}{E_1 t_1} + \frac{1}{E_2 t_2} \right], \tag{4.2}$$

which is a function of the shear modulus G_A and thickness t_A of the adhesive, as well as the Young's moduli E_1, E_2 and thicknesses t_1, t_2 of the adherends 1 (strap) and 2 (lap), respectively. With the data for the adhesive's material properties from Löbel [Löb16, p. 75], it follows

[1] Hexcel, Hexply 8552/IM7 (12K)

[2] Henkel, LOCTITE EA 9695 050NW AERO also known as HYSOL EA 9695 .050 PSF NW

$$\beta^2_{\mathrm{CLS}} = \frac{900.98\,\mathrm{N\,mm^{-2}}}{0.1\,\mathrm{mm}} \left[\frac{2}{63{,}829.4\,\mathrm{N\,mm^{-2}} \cdot 2\,\mathrm{mm}}\right] \approx 0.141\,\mathrm{mm^{-2}} \tag{4.3}$$

for the estimate, which yields

$$L_{\mathrm{CLS_{min}}} = \frac{3}{\sqrt{0.141\,\mathrm{mm^{-2}}}} \approx 7.98\,\mathrm{mm} \tag{4.4}$$

for the approximate load transfer length of the CLS specimens used in this thesis. The Young's moduli of the adherends are calculated from the 8552/IM7 properties via the classical laminate theory[3] using the parameters collected for the prepreg material by Völkerink [Völ22, p. 270].

Figure 4.2 illustrates the design of the two-stage MDAF based on the load transfer model. It shows the normalised strain in the strap for the crack lengths of 10 mm and 35 mm over the first 65 mm of the CLS specimen's overlap. The 10 mm crack length represents the initial state due to the artificial disbond of the CLS specimen. The 35 mm crack length corresponds to a crack that has reached the end of the first DAF instead (see Fig. 4.1 for the dimensions of the CLS specimen).

The strain is maximal at the crack tip before decreasing to a plateau of half the maximal value due to the load transfer to the lap thereafter. This is because no load can be transferred to the lap directly at the crack tip. As a result, the complete load is carried by the strap at this position, whereas the lap is unloaded. As the distance from the crack tip increases, the load in the strap decreases while it increases in the lap until it is equalised in both parts. Figure 4.2 demonstrates that the required distance for equalising the load is the load transfer length, approximately. Moreover, the region of the load transfer moves with the crack tip. Hence, the crack growth alters the stress/strain distribution in the CLS specimen. Therefore, it is feasible to define the sensor positions for damage detection using the load transfer model.

The clue in the two-staged MDAF design is implementing distances between the first DAF and the initial crack tip and in-between its DAFs that exceed the approximated load transfer length $L_{\mathrm{CLS_{min}}}$ by a few millimetres. The solid green curve (10 mm crack length) in Fig. 4.2 indicates that both DAFs lie in the strain plateau by placing the first DAF 15 mm away from the artificial disbond of the CLS specimen. The dashed orange curve (35 mm crack length) further demonstrates how the distance of 10 mm between both DAFs ensures the second DAF remains in the strain plateau, even if the crack has reached the end of the first DAF. As a result, both strain values are expected to be the same as long as the distance between the crack tip and the first DAF is larger than the load transfer length. With increasing crack length, however, the strain value in the first DAF sensor starts rising when the crack tip approaches it, whereas the strain value of the second DAF sensor remains at its low level.

[3] All calculations have been made in the eLamX 2.6 software [Hau23]

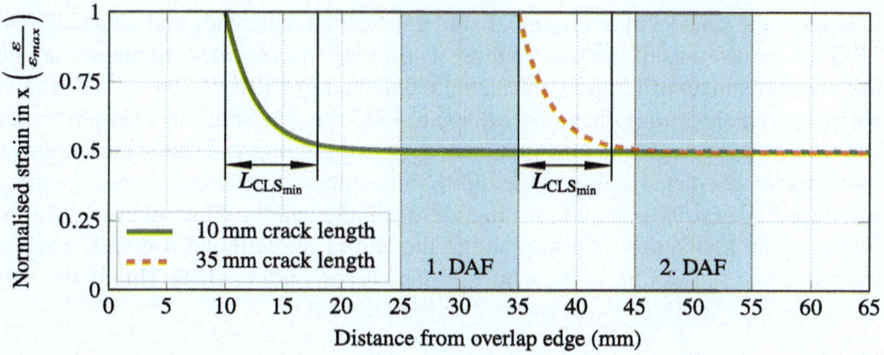

Fig. 4.2 Illustration of the DAF positions based on the load transfer model. The curves depict the normalised strain in the x-direction in the strap of a CLS specimen

The load transfer model is valid for adhesives deforming elastically only. For determining the distance between the surface toughening elements in the CLS specimen, this model suffices. The simulated strain distributions for the hybrid bondline for different crack lengths also confirm that 10 mm are a sufficient distance for the strain to decrease from the peak to the plateau value (see Fig. 3.6).

In theory, a difference in both strain values indicates a damaged bondline section in this configuration. However, the proposed damage indication method relies on the assumption that there is no dominant strain gradient due to the applied load except for the load transfer mechanism of the doubler joint. For the CLS tests in this thesis, this is true, as the tensional load is dominant, even though there is some bending due to load eccentricities. This has to be kept in mind when implementing this sensor concept in practical applications that might experience different load types. To study the fundamentals of the novel multifunctional bondline in CLS tests, the two-stage design is suited well.

In general, the two-stage design of the MDAF has another advantage. In case the crack passes the first DAF, the second DAF is capable of arresting the crack again. Hence, there is redundancy and therefore additional damage tolerance in a single two-stage MDAF. Moreover, the design reduces the risk that weak bonds might cause immediate failure of the two-stage MDAF because the weak bond had to spread across both DAFs for that.

Multifunctional Disbond Arrest Feature with FBG Sensor Arrays

The first two-stage MDAF design is made of two surface toughening elements for disbond arrest and an integrated FBG sensor array[4] for damage detection. The FBG sensor array is co-cured to the strap on top of the surface toughening elements. Hence, the sensors are placed directly in the interface of the strap and the epoxy adhesive after bonding. By leading the fibre out of the strap in front of the overlap edge, it is possible

---

[4] Miopas, FBG Arrays

to contact the sensors at a minimal disturbance of the adherends and adhesive. The FBG sensor array itself consists of three 3 mm long discrete measurement gratings, has a fibre diameter of 125 μm, and an outer diameter of 150 μm. The wavelengths of the measurement gratings in a single fibre are 1535 nm, 1545 nm and 1555 nm, which allows separating the strain information of the three gratings in a single measurement spectrum. As depicted in Fig. 4.1a and b, the centres of the measurement gratings are placed 10 mm, 30 mm and 50 mm away from the overlap edge. Accordingly, the centre of the first sensor is positioned at the end of the artificial disbond, whereas the other two sensors lie in the centre of the corresponding DAF. This is the only two-stage MDAF design with a sensor outside the DAFs.

Multifunctional Disbond Arrest Feature with Smart Inlays

Smart inlays, which functionally integrate the disbond arrest and SHM functions, form an alternative two-stage MDAF. The smart inlays are foil sensors with multiple micro strain gauges. Three generations of the smart inlay contribute to the functional compliance investigations of this thesis. The disbond arrest function is provided either by the substrate itself (first generation) or by a PVDF cover (second and third generation). The presented smart inlay designs are the result of close cooperation with the TU Braunschweig Institute of Mircotechnology, particularly with Chresten von der Heide, who is mainly responsible for the sensor layout and manufacturing of the smart inlays [Hei+21, p. 5; Hei+22, p. 3].

Figure 4.3 depicts the layout of the first- and second-generation smart inlays. Both consist of two 10 mm wide foil sections, which form the DAF as well. The half bridges are labelled according to their position in the CLS specimen after integration of the smart inlay. The letters L, M, and R stand for the half bridges positioned at the left, middle, and right of the DAFs, respectively. The number indicates whether the half bridge is in the first or second DAF. The half bridges labelled with number 1 are in the DAF 25 mm away from the overlap edge, and those labelled with number 2 in the DAF 45 mm away from the overlap edge. The centre of the sensor grids in the x-direction lies approximately 2 mm away from the edge of the DAFs, i.e., at 27 mm and 47 mm distance from the overlap edge of the CLS specimen. Figure 4.3 shows the labelling for the first DAF.

In addition to the two DAFs with integrated sensors, the smart inlay consists of a track bridge and a connector pad. As depicted in Fig. 4.1c, on the one hand, this allows for connecting the sensors outside the overlap edge without the need to cut or distort the fibres of the adherends. On the other hand, placing the track bridge in the crack growth area of the CLS specimen might influence the crack growth behaviour. Moreover, it means the track bridge is subjected to the elevated strains around the crack tip.

The main differences between the first- and second-generation smart inlays are the stacking sequence and the material selection. As depicted in Fig. 4.4 the first-generation smart inlay consists of a 100 μm thick PVDF substrate to carry the sensor structures. As a result, the sensor structures are in direct contact with the epoxy adhesive after bonding. Moreover, this means the first-generation smart inlay comes

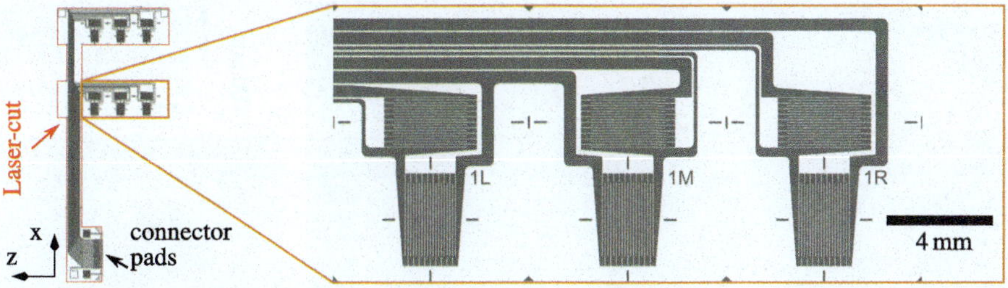

Fig. 4.3 Sketch of the first- and second-generation smart inlay layout including a detailed view of the first sensor row. Previously published in [Hei+21, p. 4]

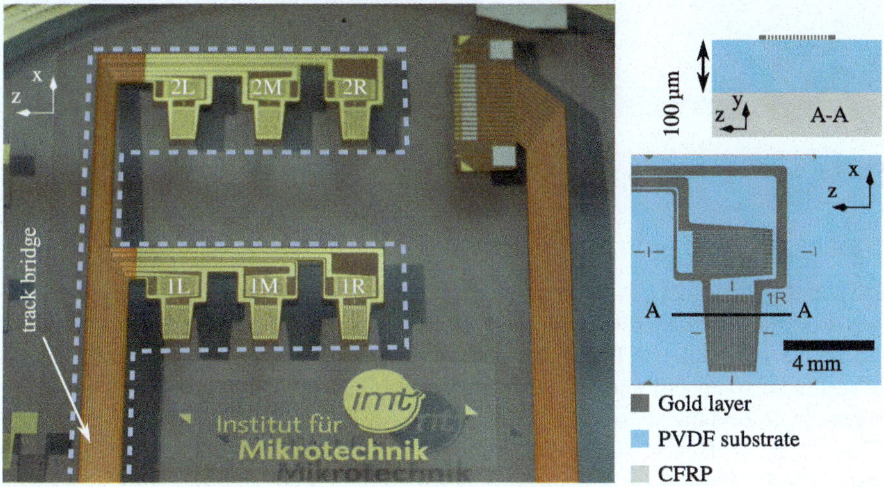

Fig. 4.4 Stacking sequence of the first-generation smart inlay

with a minimal extra material input as only the sensor structures, including the track bridges, are added in comparison to a bondline with surface toughening elements only.

As depicted in Fig. 4.5, the second-generation smart inlay differs from the first one by changing the sensor substrate to PEI and placing the PVDF as a cover on top of the sensor structures [Hei+22, p. 6]. The purpose behind this modification is to investigate whether the change in the substrate and covering the filigree sensor grids with PVDF rather than exposing it directly to the bondline improves their durability.

As explained in detail in Sect. 4.3.1, replacing the PVDF substrate is necessary for the encapsulation of the sensor because a melting substrate and cover causes severe damage to the sensor structures upon manufacturing.

PEI does not melt in the manufacturing process, but due to the combination of two different thermoplastics, a new interface is created. This is an additional potential source of error. Since it is known that both PVDF and PEI show good adhesion to the

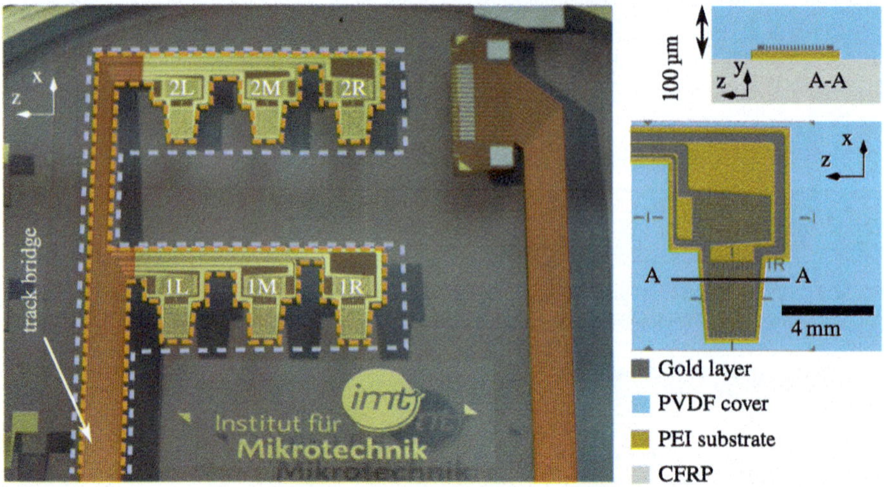

Fig. 4.5 Stacking sequence of the second- and third-generation smart inlay. Previously published in [Hei+22, p. 6]

8552 matrix [Kyr+21], whereas the adhesion between PEI and PVDF is unknown, two different types are tested. In the first type, the PEI foil separates the matrix and the PVDF completely. Hence, the functionality of the DAF relies entirely on the adhesion between PEI and PVDF. In the second variant, the PEI substrate is cut out precisely around the sensor structures (Fig. 4.5) and roughened by a laser treatment [Hei+22, p. 3]. Hence, some PVDF is still in direct contact with the matrix. As a result, the DAF relies on the adhesion between the PVDF and matrix as well as the adhesion between PEI and PVDF.

The third-generation smart inlay uses the same stacking sequence as the second generation, but modifies the concept for positioning the sensor grids and the track bridge on the substrate. Figure 4.6 depicts the modifications. The distance of the sensors to the edge of the substrate facing the crack growth area is increased by 2 mm (Fig. 4.6a). The purpose to reduce the strain at the sensor after crack arrest by moving it further away from the stress peak at the crack tip. Moreover, the connector pad is placed behind the second DAF and not in front of the specimen's overlap edge. In contrast to the first two generations, the crack growth area in front of the two-stage MDAF remains undisturbed, and the track bridges are not subjected to stress peak around the crack tip either. However, a pocket has to be milled into the lap for accessing the connector pad, which is reduced by about a third of its width around the connector pad (Fig. 4.6b). This is expected to disturb the load distribution around the pocket significantly. The purpose of this design is to move the track bridges out of the crack growth area to study how this influences the durability of the micro sensor structures.

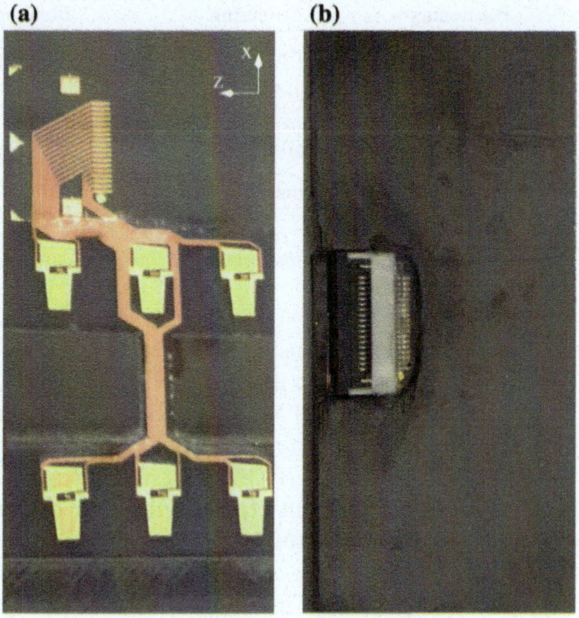

Fig. 4.6 Layout of the third-generation smart inlay **a** Smart inlay layout **b** CLS specimen with pocket milled into the lap to enable connecting the smart inlay

4.1.3 Cracked Lap Shear Specimen Manufacturing

How are the CLS specimens manufactured?

The challenges for the functionally compliant manufacturing of the multifunctional bondline come into view when looking at the standard processes for curing and bonding of the CFRP adherends. As mentioned in the previous section, the adherends of the CLS specimens are made of a unidirectional prepreg and bonded with an epoxy film adhesive. Both, curing of the adherends and adhesive bonding are autoclave processes, which means the integrated sensors are subjected to elevated temperatures and pressures. Testing the sensor integrity after these processes shall provide the answer if it is feasible to integrate the MDAFs, consisting of embedded sensors, using standard processes. As explained in Sect. 4.1.7, the feasibility of providing the piezoelectric effect of polarised PVDF is tested with a different method.

The manufacturing of the CLS specimens is a multi-stage process. Figure 4.7 illustrates the main manufacturing steps for specimens containing smart inlay MDAFs of the first or second generation, but the steps are similar for all MDAFs with embedded sensors. The first step is positioning the smart inlays on a strap plate and PVDF inlays on a lap plate. In the second step, the MDAFs are co-cured to the CFRP plates in an autoclave process. Then, the cured plates are bonded in a second autoclave process, before they are cut into single specimens. The smart inlay MDAFs are equipped with a Zero Insertion Force (ZIF) connector in a Surface Mount Technology (SMT)

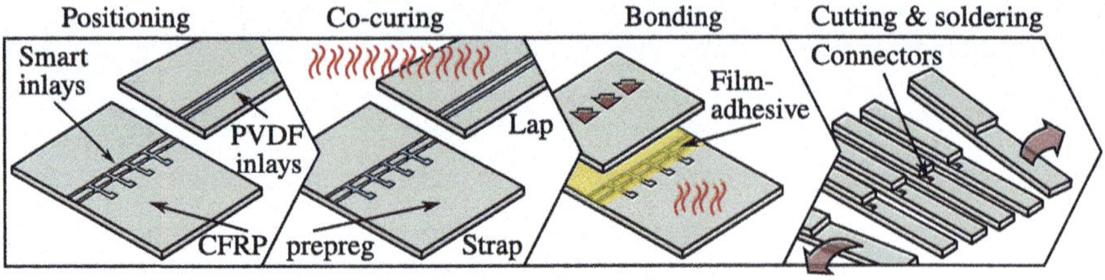

Fig. 4.7 CLS specimen manufacturing with integrated smart inlays

reflow soldering process using a low-temperature solder paste[5] to avoid melting of the PVDF material during this process.

In the case of the FBG MDAFs, the FBG fibres are positioned on top of PVDF inlays in the strap. In front of the overlap edge, the FBG fibres are placed in Poly(tetrafluoroethylene) (PTFE) tubes and on a release film. This enables pulling parts of the FBG fibre out of the CFRP after co-curing for contacting it. During co-curing, a sealant covered by a release film supports the FBG fibres that need to be led out through the side of the specimen. This is crucial for ensuring the fibres do not break during curing. Figure 4.8 depicts the positioning of the FBG fibres before co-curing.

As the co-curing and adhesive bonding processes are most demanding for the filigree sensor structures, they are briefly explained in more detail in the following. Figure 4.9 illustrates the setup and autoclave process for co-curing the smart and PVDF inlays to the prepreg. The caul plate ensures the inlays are pressed into the CFRP, which is necessary to obtain the desired smooth surface after curing. This setup has been developed by Löbel [Löb16, pp. 79–81] for co-curing the thermoplastic strips of the hybrid bondline. As indicated in Fig. 4.9b, the maximum temperature during curing is 180 °C. Since the melting point of the PVDF material is approximately 163 °C, the PVDF melts during the co-curing process. This has a major effect on the intrinsic sensor and the smart inlay concept. For the former, it has to be clarified if it is possible to maintain or recover the piezoelectric effect after co-curing of the PVDF material. For the latter, the main challenge is ensuring the integrity of the sensor grids while the substrate is melted.

The bonding process is not only demanding in terms of the applied pressure and temperature during autoclaving but also the necessary surface pretreatments. The surface pretreatment consists of cleaning the surfaces thoroughly with isopropanol and an atmospheric pressure plasma treatment using a plasma generator[6] with an RD 1004 nozzle. The parameters for the plasma treatment come from an experimental parameter study of Löbel [Löb16, pp. 107–118]. In agreement with his results, the chosen setup is a conical plasma jet head with concentric 4 mm vent, a nozzle distance of 15 mm, and a row pitch of 3 mm. During the plasma treatment, the sensor grids are

[5] Almit, LFM-65W A75C(L) 11 %.

[6] Plasmatreat, FG 5001.

Fig. 4.8 Positioning of FBG fibres before co-curing **a** Unprotected FBG fibres on top of PVDF DAFs (left side of **a**). FBG fibres placed in a PTFE tube and on a release film to enable pulling the fibres in this area out of the CFRP after co-curing (right side of **a**) **b** FBG fibres led out through the side of the plate supported by a sealant covered by a release film

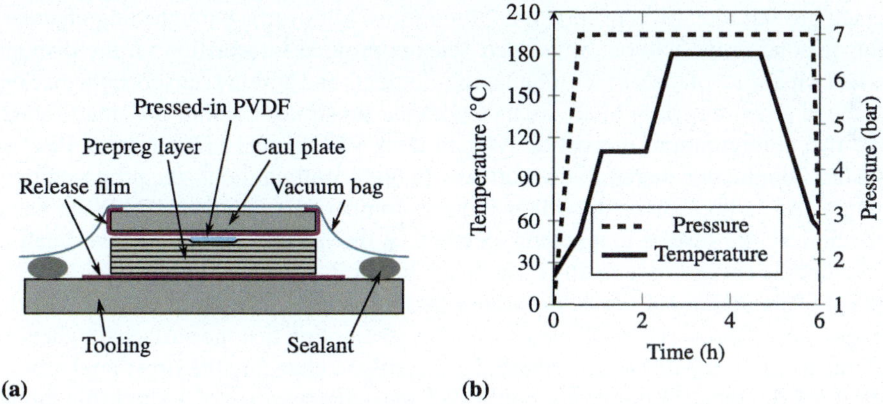

Fig. 4.9 Co-curing process **a** Vacuum setup **b** Autoclave process

directly subjected to the plasma beam in the case of the first-generation smart inlay MDAFs. In contrast, the sensor structures of the second- and third-generation smart inlay MDAFs are only indirectly subjected to the beam through the PVDF layer on top of them. The FBG sensor fibre is also not directly subjected to the plasma, as the fibre coating surrounds it. After the plasma treatment, it follows bonding of the strap and lap plates with the film adhesive. A release foil placed at the first ten millimetres between the strap plate and the adhesive creates the artificial disbond. Alignment of the smart inlays in the strap and PVDF inlays in the lap is ensured by fixing the edge of the plate with metal pins. The setup is then placed in a vacuum bag for curing the epoxy adhesive in an autoclave at 130 °C and 3 bar for two hours.

The manufacturing of the smart inlays themselves is challenging, too. This is the focus of the work of Chresten von der Heide. A detailed description of the manufacturing process of the first- and second-generation smart inlays is provided in [Hei+21, pp. 5–6] and [Hei+22, p. 3], respectively. Regarding the smart inlay manufacturing, the processes of the second and third generations are basically identical.

4.1.4 Crack Growth Tests of Cracked Lap Shear Specimens

What does the crack growth test look like?

The crack growth test is the core method for evaluating the functional compliance of the proposed MDAFs. This section explains the specimen selection and test setup.

Seven different CLS specimen configurations are tested in the scope of this thesis. Four of them are the MDAFs with embedded FBG sensors and the three smart inlay generations illustrated in Fig. 4.1 of Sect. 4.1.2. In addition, a reference configuration with neither sensors nor DAFs is tested to study the influence of the MDAFs on the crack growth rate. Moreover, two configurations with surface toughening elements for disbond arrest without embedded sensors provide information on the disbond arrest capability of the surface toughening elements and how the sensors influence the disbond arrest function. In accordance with the substrates used for the smart inlays, the two configurations consist of 100 μm-thick PVDF[7] and 125 μm-thick PEI[8] as surface toughening material, respectively. In both configurations, the proposed two-stage DAF is implemented to allow a direct comparison with the MDAFs. In some specimens, the surface toughening elements at the lap are left out for investigating whether the surface toughening elements in the strap alone are sufficient for disbond arrest. This additional test is motivated by the fact that the lap is unloaded at the beginning of the overlap, which leads to the assumption that the surface toughening elements in the lap do not contribute significantly to changing the stress field around the crack tip. If this assumption is confirmed, the manufacturing of the multifunctional bondline could be further simplified because it would eliminate the need to align the surface toughening elements in the lap and strap.

Experimental Setup

The crack growth measurements consist of 10^6 sinusoidal load cycles at a frequency of 8 Hz and three different load levels. This procedure is derived from crack growth experiments conducted by Löbel et al. [Löb+15]. All crack growth tests are carried out load-controlled at room temperature. The ambient temperature and humidity are neither controlled nor tracked. This approach seems reasonable as long as the aim is to study the fundamental working principles of the multifunctional bondline. The number of cycles is in agreement with common test standards for fatigue of adhesive lap joints [AM21, p. 322]. The test frequency is a good trade-off between

[7] Nowofol, PVDFCU-0100-0500-natu8001-0, Nowoflon PVDF, Copo FL2608.

[8] Vink König Deutschland GmbH, Ultem 1000.

testing time and the frequency-dependent stiffening for adhesives of type EA9695, as experiments of Löbel et al. [Löb+15, p. 4] with similar specimens demonstrate.

Three load levels are tested, whose load maxima are based on limit and ultimate strains for composite structures induced in the strap outside the overlap. As 3000 μm m^{-1} is the maximum of a typical range of design limit strains in composite materials [BS16, p. 370], the according load level is picked as the lower boundary for the load maxima in the crack growth tests. The upper boundary is a maximum strap strain of 4000 μm m^{-1}, which is a common ultimate strain level for composites [BS16, p. 416]. An interim strap strain of 3500 μm m^{-1} completes the load level selection. Table 4.1 provides an overview of the parameters for each load level. Again, this procedure is in agreement with the CLS tests of the hybrid bondline conducted by Löbel [Löb16, pp. 128–136].

During sinusoidal loading, the specimens are kept within the tensile regime throughout the entire load cycle, with a load ratio

$$R = \frac{P_{\text{min}}}{P_{\text{max}}} = 0.1 \tag{4.5}$$

between the maximal load P_{max} and the minimal one P_{min}.

There are three settling cycles before sinusoidal loading. In each settling cycle, the load increases to the corresponding mean load within ten seconds, stays there for five seconds and decreases again in ten seconds. Three images of the specimen are taken in the unloaded state and at mean load each before the cyclic loading starts. Moreover, the cyclic loading is interrupted once a minute (or every 480 cycles) to take another three images at mean load to enable crack growth and strain measurements through DIC. Figure 4.10 illustrates the entire test sequence, including the optical measurements.

The crack growth tests are carried out in a tensile test rig[9] equipped with a Digital-single-lens Reflex (DSLR) camera[10] for tracking the crack growth and DIC measurements as depicted in Fig. 4.11a. Figure 4.11b shows the image of a specimen recorded

Table 4.1 Load levels for the cyclic testing of the CLS specimens for all strap strain maximums

Strap strain maximum	P_{min}	P_{max}	P_{mean}	P_{amp}
3000 μm m^{-1}	0.928 kN	9.28 kN	5.104 kN	4.176 kN
3500 μm m^{-1}	1.086 kN	10.86 kN	5.973 kN	4.887 kN
4000 μm m^{-1}	1.243 kN	12.43 kN	6.8365 kN	5.5935 kN

[9] Instron, 8800 OR Zwick-Roell, Amsler HC25. The test rig was changed from the Instron 8800 to the preferred Zwick HC25 in the middle of the test campaign.

[10] Canon EOS 5D Mark IV with Zeiss Milvus 2/100M lens.

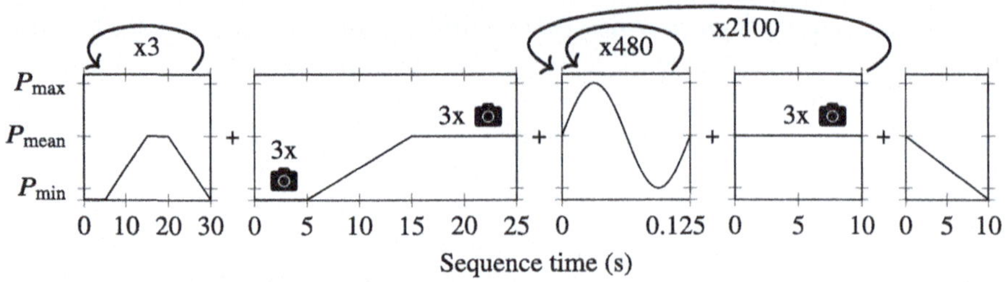

Fig. 4.10 Crack growth test sequence

Fig. 4.11 Crack growth measurement test set up **a** Test rig and the camera setup for crack length and DIC measurements **b** CLS specimen recorded by the camera

during testing. With a few exceptions, a minimum of three CLS specimens of each type are tested at each load level. Table 4.2 provides an overview of the number of specimens tested for each configuration and load level.

A multichannel strain gauge amplifier[11] in combination with commercial software[12] records the load and displacement values, which are provided by the internal load cell[13] and displacement sensor[14] of the test rig, as well as the outputs of the smart inlays. The first ten seconds of each loading interval are recorded at a high sampling frequency of 150 Hz. In addition, peak values are stored for every cycle.

[11] HBM, QuantumX MX1616B.

[12] HBM, Catman Easy.

[13] Huppert, 1010-BPS-25kN-5/8".

[14] RDP, LVDT ACT3000.

Table 4.2 Number of CLS specimens tested in the crack growth experiments for each configuration and strap strain maximum

	Strap strain maximum		
CLS configuration	3000 μm m^{-1}	3500 μm m^{-1}	4000 μm m^{-1}
3rd-gen. smart inlay MDAF	5	3	4
2nd-gen. smart inlay MDAF	5	3	1
1st-gen. smart inlay MDAF	6	4	4
FBG MDAF	3	3	3
PEI DAF	2	0	1
PVDF DAF	6	4	3
No DAF	6	6	3

The transformation of the smart inlay outputs from mV V^{-1} to μm m^{-1} is based on the calibration of an extra specimen in a tensile test [Hei+22, p. 4]. This specimen consists of smart inlays co-cured to a strap without a lap bonded to it. According to the calibration, 1.02 mV V^{-1} equal 1000 μm m^{-1} [Hei+22, p. 10]. The according ratio is applied to convert all smart inlay outputs of the second and third generation to micro strain.

An interrogator and its corresponding software tool[15] measure the spectra of the FBG sensors and detects the wavelengths of the measurement gratings via an internal peak detection algorithm. As the software only offers continuous measurements, the FBG data is recorded with a sampling frequency of 80 Hz. According to Prussak et al. [Pru+18, p. 261], the strain of the FBG sensors

$$\varepsilon_p = \frac{\frac{\lambda_b - \lambda_0}{\lambda_b} - C_T \times \Delta T}{C_s} \tag{4.6}$$

is a function of the measured Bragg wavelength λ_b, the Bragg wavelength of the unloaded state λ_0, the coefficient of strain C_s, which is a constant of 0.78, the temperature difference ΔT, and the thermo-optic coefficient C_T. Since the temperature difference at the FBG sensor is not tracked, it is assumed to be zero. Moreover, λ_0 is determined after the three settling cycles.

The optical measurements of the crack growth and strain are conducted on one side of the CLS specimen by the DSLR camera. A red speckle pattern on a white background is applied to the side of the specimens facing the camera. Since a crack appears as a black object in the image, using a red speckle pattern allows for measuring the crack length and strain measurements from the same image.

[15] Miopas, SPK 155.

Crack Length Evaluation

The crack length measurements require a tool to separate the black crack from the red speckles in the recorded images. The binarisation of an image containing a specimen with a crack is not trivial. A simple threshold is insufficient since the size of the crack decreases towards its end. This means fewer black pixels and more overlap with the material. This can lead to the removal of pixels that semantically belong to the crack, which could distort the measured length of the crack. For that reason, a new algorithm called "Island Hopper" developed by Rashid Kratou and Jan-Christopher Schmidt of the Gauß-IT-Zentrum at the TU Braunschweig is used.[16]

The basic working principle of the Island Hopper algorithm is illustrated in Fig. 4.12. The first step is always cropping the original image around the crack growth area (Fig. 4.12a). This is necessary to remove all artefacts that are not part of the specimen. The algorithm starts with preprocessing the cropped images. This includes the following sequential steps: converting the image into the HSV (Hue Saturation Value) space, extracting the Value channel as brightness of the pixels, inverting the Value channel, and enhancing its contrast via a quadratic function. The contrast-enhanced image is then Sobel-filtered to detect edges. All pixels that are not on a Sobel edge with a minimum value are set to two-thirds of the maximum brightness of the contrast-enhanced image, provided their values exceed that threshold. Pixels on a Sobel edge above the threshold retain their values. Finally, the pixels are divided by the maximum value of the image. The result is an 8-bit image containing the edges of the original image and all pixels from the Value channel with values in the contrast-enhanced image that are higher than a relative threshold; the retained pixels from the Value channel are set to this threshold value. These are the candidates for the following Island Hopper algorithm. This image is referred to as the Sobel image. The Sobel image is binarised, which yields an image in which the crack and some remaining speckles appear in white (Fig. 4.12b).

Then the Island Hopper algorithm starts using the enhanced-contrast and the Sobel image. A Connected Components algorithm clusters connected pixels to "islands". A Hough Line Transform is used to get a line through the islands (blue line in Fig. 4.12c). For each of these islands, an island quality value is calculated, based on the number of pixels of this island, the brightness of these pixels in the Sobel image and contrast-enhanced image, and the distance from the Hough line. Thereafter, the algorithm assigns islands to the crack. The island with the highest quality is the first to be assigned. The algorithm then assigns the island with the next best relative quality to the crack. The relative quality is a second value determined for each island based on the distance to the last island that has been assigned to the crack and the island's own qualities. Islands with a higher distance from the last island are ranked lower. The distance function adds a penalty for distance in the y-direction, to account for an assumed crack propagation in the x-direction. This is continued until there is no island left whose quality reaches a threshold which has been determined by trial and error. Afterwards, a second crack line is drawn through the last islands of the

[16] The description of the working principles of the "Island Hopper algorithm" in the following two paragraphs is based on a text written by Rashid Kratou and revised by the author.

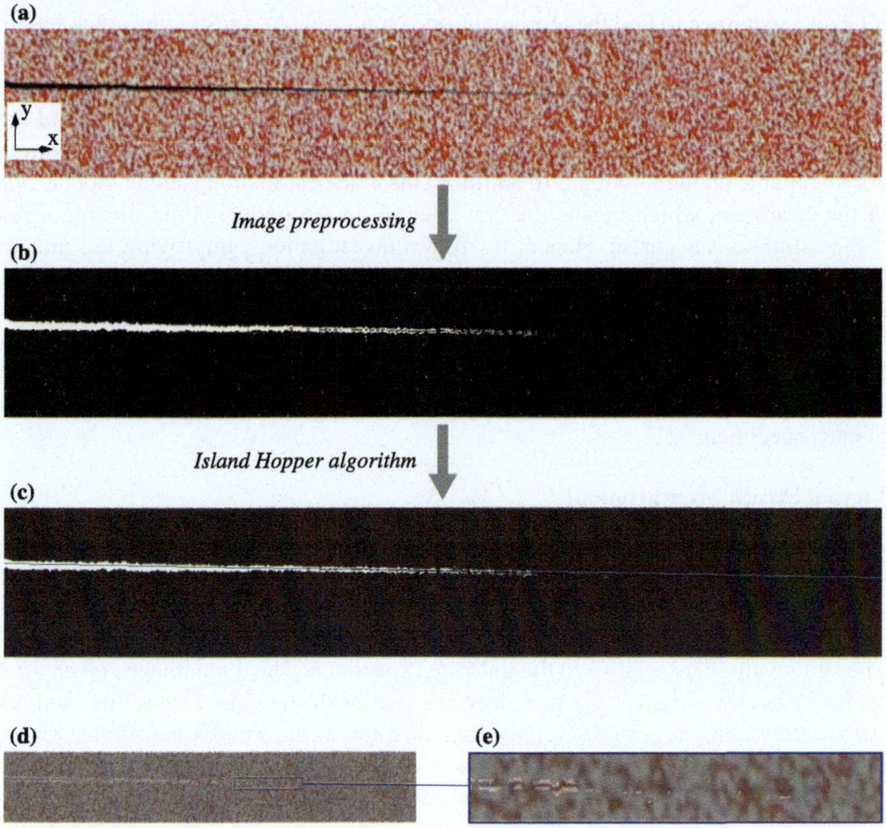

Fig. 4.12 Working principle of the Island Hopper algorithm with example images **a** Cropped input image **b** Binary image after preprocessing **c** Binary image after application of the Island Hopper algorithm with the crack tip as the most right white pixel. Crack line in blue **d** Result of the Island Hopper algorithm laid over the cropped image **e** Zoom on the obtained crack tip

crack. If there are islands very close to that line, these are also assigned to the crack if they are within a predefined distance of islands associated with the crack. This ensures small islands at the crack tip are not missed. The most right white pixel of all considered islands determines the crack tip (Fig. 4.12c).

Despite the advanced algorithm, the crack length measurement is not completely accurate. A systematic problem appears to be the tiny crack opening, which causes the islands at the end of the crack to get rather bad qualities for their size (see Fig. 4.12c and e). Since the filtering cannot remove all speckles without removing parts of the crack tip, a trade-off is often necessary between allowing too many islands that do not belong to the crack and removing too many that are part of the crack. This is especially true for some specimens, whose image quality is poor as well. Moreover, the parameters for the algorithm, in particular the preprocessing, require adjustment for every specimen to find the aforementioned trade-off. As a result, the crack length estimations scatter, and are often too short in comparison to a manual evaluation of crack length reading from the images. The crack length estimate still allows for a qualitative evaluation regarding the correlation between the sensor outputs and the crack length. For a precise quantitive evaluation of the crack length, the results might not be reliable enough, though. In addition, the crack estimation is done on one side of the specimen, which means the crack length in the middle could diverge, even if the estimate is accurate. Hence, for future investigations, improving the images for a clearer separation between the speckles and the crack front is as important as measuring the crack length inside the specimen for a precise quantitive correlation between the sensor output and the crack length. Nevertheless, the crack estimates enable the correlation between the sensor data and the crack length. The evaluation of the disbond arrest function, however, relies on the manually determined crack tips of each specimen.

Virtual Strain Measurement

The speckle pattern on the CLS specimens enables virtual strain measurements in addition to the integrated sensors. For the virtual strain measurement, the red colour channel of the RGB image is subtracted to increase the contrast between the speckles and the white primer. The strains are then obtained by a commercial DIC software.[17] The strain values are relative to the unloaded specimen. The three images taken after each 480 cycles are averaged to reduce the scatter of the data. Using this method, six virtual strain gauges (VSGs) are implemented at the same x-coordinate as the FBG sensors. In contrast to the FBG sensors, three VSGs each measure the strain in the lap and the strap, approximately one millimetre away from the bondline in the y-direction (see Fig. 4.13 for the positioning). This allows for comparing the strains obtained in the bondline from those in the adherends. Moreover, the VSGs average the strain over three millimetres for obtaining a similar behaviour as the FBG sensors.

4.1.5 Quasi-Static Tests of Cracked Lap Shear Specimens with Varying Artificial Disbond Lengths

For the evaluation of the damage detection capability of the multifunctional bondline, quasi-static tests of CLS specimens with three different artificial disbond lengths support the findings from the crack growth tests. The advantage of these tests over the crack growth ones is the well-defined disbond length. This allows for an unambiguous correlation of the disbond length to the sensor outputs, which cannot be guaranteed in the crack growth tests because the crack length is only measured on the side of the specimens. This section presents the method for these tests.

In the quasi-static test, only two different MDAF types are tested, i.e. first- and second-generation MDAFs. For each type, the test includes a single specimen for

[17] GOM Correlate.

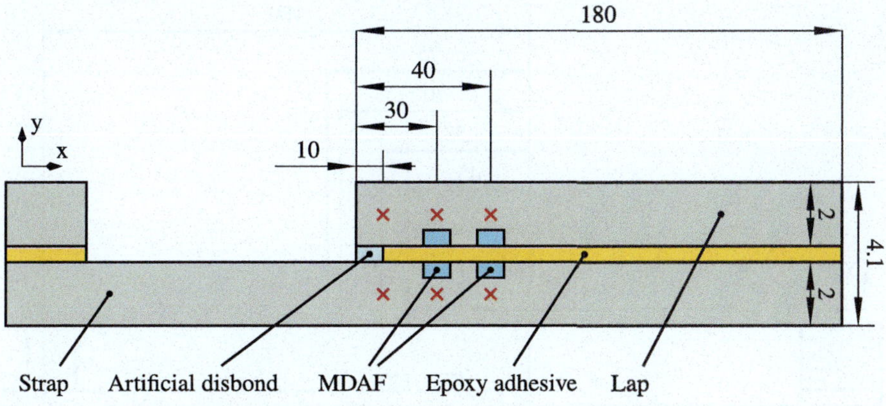

Fig. 4.13 Side view of the CLS specimen with VSG positions marked by an x

three different artificial disbond lengths of 10 mm, 16 mm and 23 mm. During the quasi-static tests, the load increases within ten seconds from 50 N to 5.104 kN, remains there for twenty seconds, and decreases in another ten seconds to 50 N. The maximum load is the same as the mean load of the crack growth test for the lowest load level. As for crack growth testing, three settling cycles are carried out before the actual measurement. For that matter, the loading sequence for the quasi-static tests is applied twice, once for settling and once for the measurement.

4.1.6 Residual Strength Test of Cracked Lap Shear Specimens

A residual strength test provides information on the static strength of the CLS specimen after a disbond arrest event. The residual strength is determined in a quasi-static test, which is displacement-driven at 2 mm/min until global failure of the specimen. There are two types of residual strength tests.

The first type takes unmodified CLS specimens from after the crack growth test.[18] CLS specimens without DAFs and with an increased artificial disbond length of 25 mm provide the reference. The increased artificial disbond creates an equivalent pre-damage as found in cycled CLS specimens, whose damage has been arrested at the first DAF. This is necessary because all specimens without DAF fail in the crack growth test (see Sect. 4.2.1). This test only involves specimens with no DAF, PVDF DAF, and first-generation smart inlay MDAF.

---

[18] This test has been carried out at DLR Braunschweig by J. Kosmann and M. Schollerer. Parts of the results are already published in Schollerer [Sch24, Sect. 7.6].

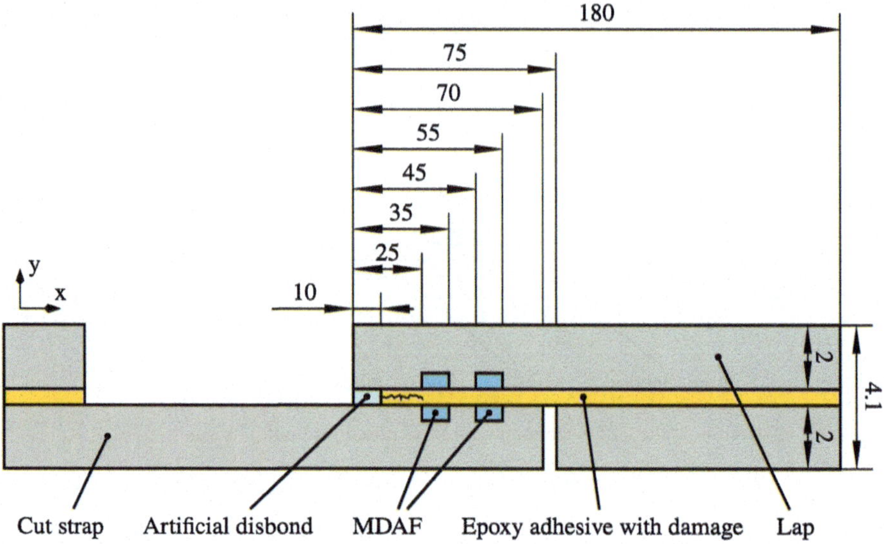

Fig. 4.14 Modified CLS specimen with cut strap and damaged adhesive for residual strength tests after the crack growth test

Table 4.3 Number of CLS specimens tested in the residual strength tests for each configuration and modification

CLS configuration	Unmodified	Modified (not cycled)
3rd-gen. smart inlay MDAF	–	5 (0)
2nd-gen smart inlay MDAF	–	5 (0)
1st-gen. smart inlay MDAF	5	11 (2)
FBG MDAF	–	12 (3)
PEI DAF	–	–
PVDF DAF	3	12 (4)
No DAF	5	5 (5)

For the second type,[19] the CLS specimen is manipulated by cutting the strap 70 mm behind the overlap edge as depicted in Fig. 4.14. The position of the cut creates a specimen comparable to SLS specimens with 70 mm overlap length and 15 mm epoxy adhesive at both sides of the MDAF. This means the specimen is symmetric if there is no crack growth.

Table 4.3 summarises the specimens tested in the unmodified and modified configurations. As there are some non-cycled specimens among the modified type, these are indicated as well. Regarding the unmodified test, all specimens are cycled except for the no DAF configuration.

[19] Tested in an Instron, 150 kN.

4.1.7 Annealing of Polarised PVDF Using Standard Curing Processes

What is the investigation method regarding the feasibility of maintaining the piezoelectric effect of PVDF throughout the curing processes?

Annealing polarised PVDF films in different autoclave cycles and measuring the phase contents before and after is the core method for investigating the feasibility of maintaining the piezoelectric phase throughout the curing processes of the multifunctional bondline. This section summarises the methods presented in [Vas+21, pp. 2–4] for these investigations.

The total crystallinity and the fraction of the piezoelectric phase are two crucial values for the piezoelectric effect of PVDF. As explained in Sect. 3.4.3 forming piezoelectric crystalline phases after bonding is not possible using conventional techniques. Hence, it is of interest to investigate how much piezoelectric active phase remains after annealing the PVDF material by using standard curing processes.

The corresponding investigation is a three-step process: The initial determination of the electroactive phase content and crystallinity of the PVDF material, annealing in different autoclave processes, and determination of the phase content and crystallinity after annealing. Obtaining the initial phase contents and crystallinity before and after annealing consists of XRD, FTIR, and DSC analyses. While the XRD and FTIR analyses do not change the phase content and crystallinity of the PVDF, the DSC does so. Hence, ten extra samples provide the initial DSC data, whereas all other samples are only characterised via XRD and FTIR before annealing in the autoclave. The autoclave process is similar to the process for curing the film adhesive, with adjustments regarding the maximum temperature and the time the temperature is applied. In particular, the heating and cooling rates of 2 K/min, and pressure of 3.0 bar are the same as for the curing process of the film adhesive. As for the maximum temperature, a total of nine different maximum temperatures have been selected for studying the impact of the annealing temperature on the phase content. Moreover, the holding time at the annealing (maximum) temperature is reduced from two hours to one hour in all annealing processes. Table 4.4 summarises the selected temperatures and explains the selection. Four PVDF samples have been annealed together in a single autoclave cycle for each maximum temperature. In the autoclave annealing process, the samples are placed between two copper plates and subjected to a vacuum of $\approx$0.3 mbar.

In contrast to the PVDF used for the DAFs, the PVDF material[20] in this investigation is only 40 μm thick and already polarised. Moreover, the area of the test samples is $15 \times 15\,\text{mm}^2$, which is slightly smaller than the surface toughening elements of the CLS specimens ($25 \times 10\,\text{mm}^2$). It is assumed that the different geometry has no significant effect on the phase transitions.

---

[20] Provided and polarised by Nowofol. The exact information on the material is classified by the manufacturer.

Table 4.4 Overview of the annealing temperatures and selection criteria for the autoclave processes according to Vasic et al. [Vas+21, p. 4] with few additions

Annealing temperature (°C)	Selection criterion
90	Typical temperature for stretching, polarising, and annealing processes of PVDF to increase the β-content
130	Curing temperature of the film adhesive in the manufacturing process of the CLS specimens
160, 165, 170, and 175	Temperatures around the typical melting temperature range of α- and β-PVDF between 167 and 172
180	Typical curing temperature of the CFRP prepreg
182	Highest annealing temperature at which Nakamura et al. [Nak+03, p. 1704] report of pure β-phase formation
185	Temperature at which Nakamura et al. [Nak+03, p. 1704] report of a formation of α-phase only

The total crystallinity

$$X_c = \frac{H_{\mathrm{m}}}{H_{100}} \tag{4.7}$$

is determined from the melt enthalpy H_m retrieved from the DSC analysis and the melt enthalpy of a 100 % crystalline PVDF film H_{100} [Vas+21, p. 5]. The value H_{100} originates from Nakagawa and Ishida [NI73, p. 2186][21] and is often taken in literature [Lop+11, p. 18080; BE03, p. 1095].

The FTIR yields the fraction of the electroactive phase

$$F_{\mathrm{EA}} = \frac{A_{\mathrm{EA}}}{\left(\frac{K_{840}}{K_{763}}\right) A_{763} + A_{\mathrm{EA}}} \times 100\,\%, \tag{4.8}$$

depending on the measured absorbance A at the corresponding wavelength and the specific wavenumbers K_{840} and K_{763} [Vas+21, p. 3]. The specific wavenumbers are literature values taken from [MLL14, p. 685]. The wavelength 763 cm^{-1} and 840 cm^{-1} belong to the α- and β-phase, respectively. The absorption bands A_{EA} include the wavelength peaks at 1275 cm^{-1} for the β-phase, 833 cm^{-1} and 1234 cm^{-1} for the γ-phase, and 840 cm^{-1} for both, the β- and γ-phases. The absorbance

$$A = -\log_{10}(\zeta), \tag{4.9}$$

is determined from the rate of transition ζ,[22] which expresses how many infrared rays a sample absorbs in comparison to reference measurements without absorbing material [Vas+21, p. 3]. The reason the fraction of the electroactive phase is illustrated

[21] Nakagawa use 25 cal g^{-1}, which equals 104.6 J g^{-1}.

[22] In [Vas+21] the transition is expressed by the variable T.

rather than the β- and γ-phases is that those phases cannot be distinguished in the FTIR spectrum. As soon as the DSC analysis shows the presence of a second or even third melting point at temperatures of 175 °C and above, the presence of the γ-phase is indicated. Therefore, the electroactive phase fractions at lower temperatures can be attributed to the β-phase, whereas it can be either β-, γ-phase or both at the higher annealing temperatures [Vas+21, p. 5].

4.2 Safe and Efficient Operation

Is the safe and efficient operation of bondlines with MDAFs feasible?

This section presents the results of the experimental investigations regarding the safe and efficient operation of the proposed multifunctional bondline. It discusses the three main aspects: disbond arrest, damage detection, and the influence of the MDAFs on the mechanical strength of the bondline. The interpretation of the results regarding the functional compliance of the MDAF is further discussed in Sect. 4.4.

4.2.1 Disbond Arrest Effect of Surface Toughening and Different Multifunctional Disbond Arrest Features

Is the surface toughening approach capable of arresting disbonds, and what is the effect of the integrated sensors on it?

This section compares the disbond arrest function of the different (M)DAF configurations by analysing the crack length at the end of the crack growth test of all CLS specimens, and the bondline damage and crack growth of selected CLS specimens.

Crack Tip Analysis

First, the crack growth results for the most crucial strap strain maximum of 3000 $\mu m\,m^{-1}$ are discussed. Figure 4.15 depicts the corresponding crack lengths obtained at the end of the CLS crack growth test for all seven specimen configurations. Each point in Fig. 4.15 illustrates the crack length of a single specimen after the crack growth test via its position on the x-axis. All crack lengths are obtained manually by inspecting the final image of each crack growth test. A crack length of 65 mm defines a completely failed bondline because the second DAF has been passed by 10 mm at this point. Some specimens consist of a DAF in the strap, but none in the lap. These are marked with an asterisk. In addition, the y-axis indicates the sample size *n* for each specimen configuration in the corresponding row. If all points of a single type fall on the same crack length, only one mark is shown.

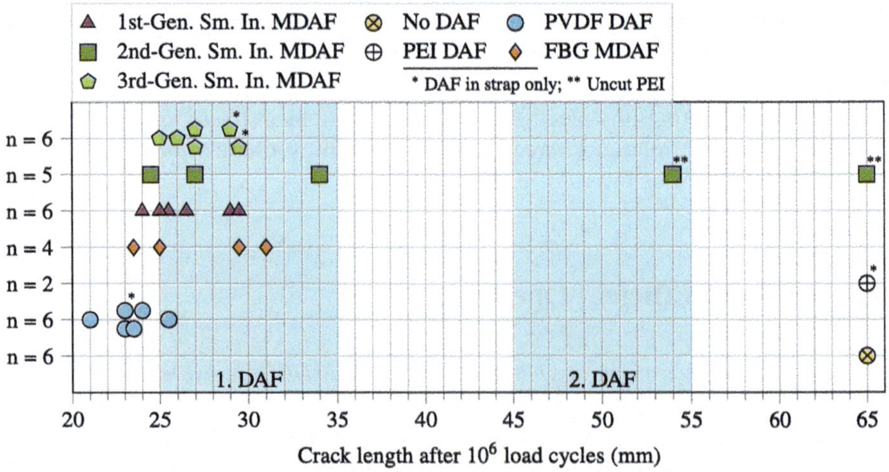

Fig. 4.15 Comparison of the crack length in various CLS specimen configurations after 10^6 load cycles at 3000 μm m^{-1} strap strain maximum. The DAFs are marked in light blue, and the sample size n for each configuration is indicated as well

As the focus is on the disbond arrest effect, the differences in the DAF configurations of all specimens are briefly repeated. In three specimens, the DAFs consist of a 100 μm-thick PVDF foil, which is in direct contact with the CFRP laminate. These are the types: PVDF DAF, FBG MDAF, and first-generation smart inlay MDAF. In the other three DAFs, a PEI foil is in contact with the laminate instead. In the case of the PEI DAF, the DAFs are made of a 125 μm-thick PEI foil. In the second-generation smart inlay MDAF, the DAFs consist of a PEI foil of $\approx$8 μm thickness as substrate, with micro strain gauges, and a 100 μm-thick PVDF foil cover. In most specimens of this type, the PEI foil is cut precisely around the sensors so that parts of the PVDF cover are in direct contact with the CFRP laminate as well (see Fig. 4.5). In some other specimens, the PVDF foil is not in contact with the laminate at all but only with the PEI. The latter are referred to as "uncut PEI" in the following and marked by a double asterisk in Fig. 4.15. All third-generation smart inlay MDAFs consist of the DAF setup with an 8 μm-thick, cut PEI substrate and 100 μm-thick PVDF cover.

All two-stage DAFs with PVDF in direct contact with the CFRP laminate stop the crack completely within the first DAF at limit strain level (3000 μm m^{-1} strap strain maximum). This includes the first-generation smart inlay MDAFs, whose strain gauges are applied on a PVDF substrate, as well as the second- and third-generation smart inlay MDAFs, whose PEI substrate is cut out closely around the sensor structures. The specimens with MDAFs show a tendency to a slightly larger crack length in comparison to those with PVDF DAFs, but all cracks are still arrested within the first DAF. Moreover, all three specimens containing DAFs based on PVDF in the strap, but none in the lap, show a disbond arrest effect. These include one specimen with PVDF DAF and two specimens with third-generation smart inlay MDAFs. In contrast, the crack grows through the entire specimen, if there is no DAF or a PEI-based DAF. In addition, the crack grows much further or is not stopped at all in the

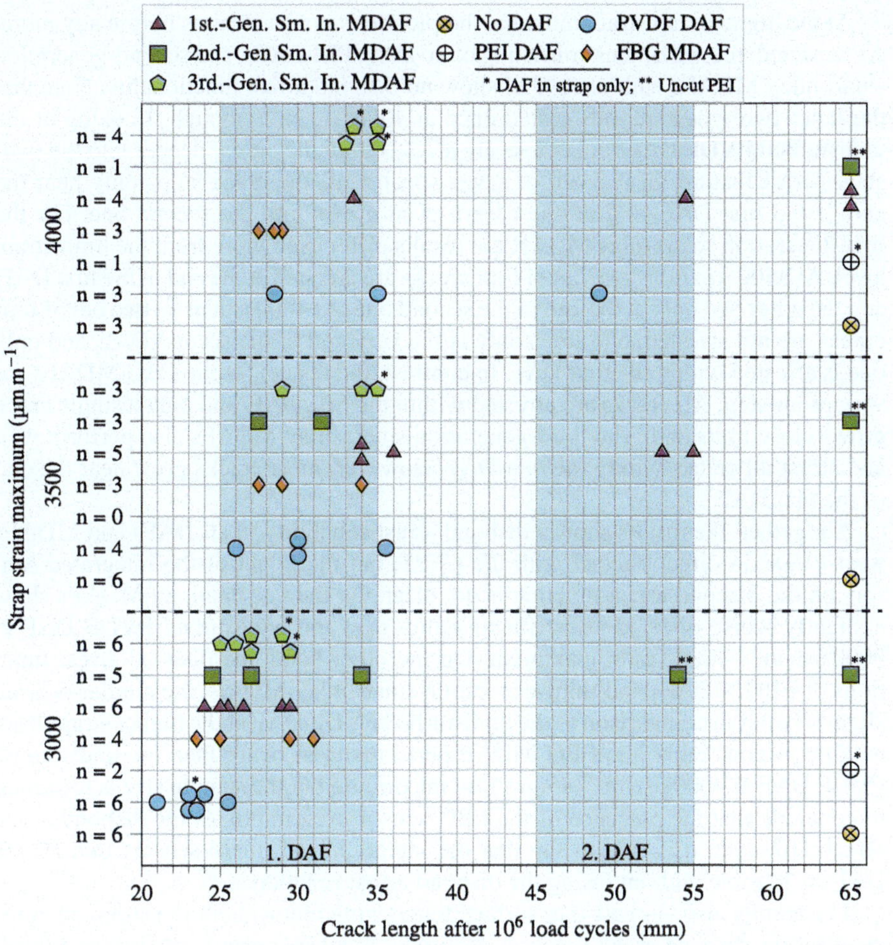

Fig. 4.16 Comparison of the crack length in various CLS specimens configurations after 10^6 load cycles at three different strap strain maximums. The DAFs are marked in light blue, and the sample size n for each configuration is indicated as well

second-generation smart inlay MDAFs with uncut PEI substrate. In conclusion, the surface toughening effect of PVDF is sufficient to arrest cracks at limit strain level, whereas PEI does not show a disbond arrest effect. In addition, the integration of FBG sensors or strain gauges into the PVDF-based DAFs is possible without jeopardising the disbond arrest capability. Placing PVDF surface toughening elements in the strap, only, is sufficient to obtain the disbond arrest function.

Figure 4.16 extends Fig. 4.15 by the results of the specimens tested at strap strain maxima of $3500\,\mu\text{m}\,\text{m}^{-1}$ and $4000\,\mu\text{m}\,\text{m}^{-1}$. The strain levels are marked on the y-axis. Other than that, the layouts of Figs. 4.16 and 4.15 are the same.

At the higher strap strain maxima, the picture is not as clear as before any more. As expected, the specimens consisting of no DAF, PEI DAFs, and second-generation smart inlay MDAFs with uncut PEI show no disbond arrest functionality. Note that the latter two types are only tested once at $4000\,\mu\mathrm{m\,m}^{-1}$, though. Looking at the undisturbed PVDF DAFs, all cracks are arrested, but they grow deeper into the two-stage arrest feature. In two out of seven specimens, the crack stops only near the end of the first DAF, and in one case it even grows into the second one. For the first-generation smart inlay MDAF, the results differ even more from the limit strain level. At $3500\,\mu\mathrm{m\,m}^{-1}$, two out of four cracks are arrested at the end of the first DAF, and the other two only at the end of the second one. At $4000\,\mu\mathrm{m\,m}^{-1}$, two out of four cracks are not arrested, one stops only at the very end of the second DAF, and only one is arrested within the first DAF. In contrast, the results for the FBG MDAFs are very promising. All cracks are arrested within the first DAF, and only a single crack grows close to its end. The third-generation smart inlay MDAFs also perform well and arrest all cracks within the first DAF. However, most cracks grow near the end of the first DAF.

Altogether, the results show a disbond arrest capability of all DAFs and MDAFs with PVDF in direct contact with the CFRP, but the effect of the integrated sensors on the disbond arrest function is not clear. On the one hand, all MDAFs show a slightly worse disbond arrest functionality than the undisturbed PVDF DAF at $3000\,\mu\mathrm{m\,m}^{-1}$ strap strain maximum. On the other hand, the disbond arrest functions of FBG MDAFs and third-generation smart inlay MDAFs are similar or even slightly better in comparison to the undisturbed PVDF DAF at the higher strap strain maxima. Remarkably, the FBG MDAF performs even better than the undisturbed PVDF DAF at $4000\,\mu\mathrm{m\,m}^{-1}$ strap strain maximum, but the sample size is too small to conclude that the presence of the FBG sensor even improves the disbond arrest effect. However, it is safe to say that the integration of micro sensors into PVDF DAFs is feasible without losing the disbond arrest function.

The results also suggest that surface toughening elements in the strap are sufficient for the disbond arrest function. However, in all specimens consisting of third-generation smart inlay MDAFs, the crack length is the longest for the variant with DAF in the strap only. This suggests that the surface toughening element in the lap might contribute slightly to the disbond arrest function, but it is not necessary to achieve it. However, the small sample size does not allow a distinct judgement regarding the contribution of the surface toughening elements in the lap to the disbond arrest function.

Bondline Damage Analysis

The small sample size in combination with the noticeable deviations within some specimen groups does not give a clear picture of the disbond arrest function at the higher strap strain levels. Thus, the bondline damage in some specimens is analysed in more detail in the following for a more profound understanding of the disbond arrest function. Moreover, this allows checking for a specific reason for the deviations.

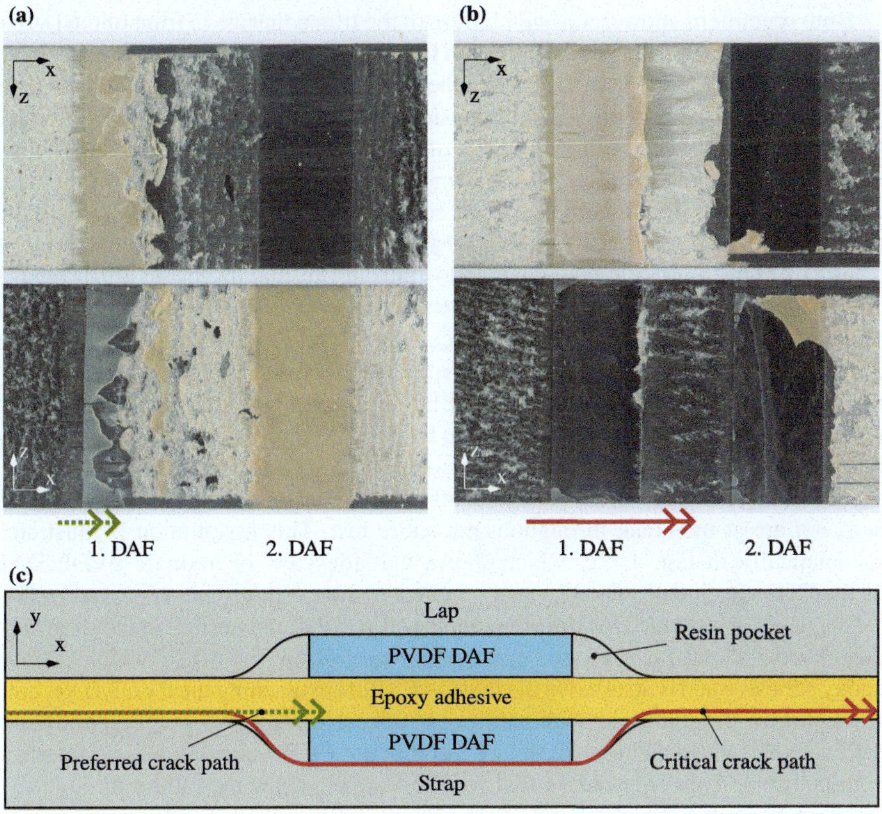

Fig. 4.17 Comparison of the bondline damage in CLS specimens with PVDF DAFs after crack growth tests at 4000 $\mu m\, m^{-1}$ strap strain maximum and consecutive quasi-static loading until global failure. The green and red arrows indicate the preferred and critical crack paths. The upper photographs are the lap and the lower the strap **a** Top view of a specimen, whose crack has been arrested in front of the 1. DAF **b** Top view of a specimen, whose crack has been arrested within the 2. DAF **c** Side view around the 1. DAF, indicating the critical and preferred crack paths

Figure 4.17 depicts the bondline damage and crack paths in CLS specimens with PVDF DAFs after the crack growth test and consecutive quasi-static loading until global bondline failure, i.e. separation of the strap and lap. The upper photographs in Figs. 4.17a and b depict the lap, and the lower photographs the strap of the CLS specimens. The maximum strap strain level of both specimens is 4000 $\mu m\, m^{-1}$. The crack tip of the specimen in Fig. 4.17a lies slightly within the first DAF. In contrast, the crack tip of the specimen in Fig. 4.17b is in the second DAF. The damage originating from the cyclic loading shows very smooth surfaces of the damaged area in the epoxy adhesive. This allows for distinguishing this damage from the one due to quasi-static loading. Compare the damages in front of the first DAF (cyclic loading) and between the DAFs (quasi-static loading) in Fig. 4.17a, for example.

Both specimens show a cohesion failure of the film adhesive in front of the DAFs. The failure mode remains a clear cohesion failure of the film adhesive in the specimen depicted in Fig. 4.17a, but in the case of the specimen shown in Fig. 4.17b it changes to either an adhesion or a special cohesion failure of the CFRP-PVDF interface at the strap.[23] Due to the transparency of the PVDF foil, it is uncertain whether the PVDF separates from the CFRP completely (adhesion failure) or a thin layer of PVDF remains on the CFRP (special cohesion failure). The adhesion or special cohesion failure occurs in the second DAFs of both specimens, too. However, this results from the quasi-static loading, and is thus not further discussed regarding the disbond arrest functions. For explaining the crack path in the CLS specimen in the scope of this thesis, it is not of importance which kind of failure it is. However, for improving the interface, it certainly is. Hence, for a clarification of the failure mode, Energy-Dispersive X-ray Spectroscopy (EDX) could help to determine if there is still a significant amount of fluoride left on the CFRP. This approach is recommended for future work.

The change in the failure mode seems to correlate with the fact that the DAF does not arrest the crack throughout the entire test. This assumption is illustrated schematically in Fig. 4.17c, which shows the side view of a single PVDF DAF of the CLS specimens. The red arrow indicates the crack growth observed in Fig. 4.17b, which passes the DAF by breaking the CFRP-PVDF interface at the strap. The green arrow shows the preferred crack path observed in Fig. 4.17a, which stays in epoxy adhesive and is arrested in the first DAF. (Before reaching the PVDF DAF, both cracks follow the green path within the epoxy adhesive.) Note that the preferred crack path does not mean a complete failure through the PVDF-epoxy adhesive interface is impossible. This only means that for the tested specimens, failure through this interface has not been observed. Moreover, it is the preferred crack path because the cohesion failure of the epoxy adhesive means a failure of a material and not an interface.

It is noteworthy that this failure mode stands in contrast to the demand of Hart-Smith [Har21, p. 786] that the failure should always grow through the adherends. Due to the presence of the DAFs, however, it seems tolerable for the failure to stay in the bondline because the DAF ensures it does not grow rapidly. In combination with the SHM function, it might even be beneficial if the failure always stays in the bondline because the damage growth in the bondline is monitored.

Failures of the CFRP-PVDF interface are observed in other specimens with PVDF-based DAFs as well. Figure 4.18a depicts a specimen with a first-generation smart inlay MDAF. The bondline fails completely during the crack growth test. The failure mode in the DAF and the track bridge of the sensor at the top of the strap is dominated by the adhesion or special cohesion failure of the PVDF-CFRP interface in the strap. In addition, there is an adhesion or a special cohesion failure of the sensor structure on top of the PVDF and the epoxy adhesive. This indicates a potentially weak gold-epoxy interface. This is further discussed in Sect. 4.2.4.

[23] The failures are named in accordance to DIN EN ISO 10365 [DD22].

Fig. 4.18 Comparison of the bondline damage in CLS specimens with smart inlay MDAFs after the crack growth test and consecutive quasi-static loading until global failure. The upper pictures are the lap and the lower the strap **a** Specimen with first-generation smart inlay, whose crack has not been arrested (Strap strain maximum of 4000 $\mu m\,m^{-1}$) **b** Specimen with second-generation smart inlay, whose crack has been arrested within the second DAF (Strap strain maximum of 3000 $\mu m\,m^{-1}$)

Figure 4.18b shows a specimen with second-generation smart inlay MDAF and cut PEI. During cyclic loading, the first DAF fails. Again, there is a failure of the CFRP-PVDF interface. The same failure is observed in the PEI-sensor-PVDF interface. In addition, the PEI sensor fails due to an adhesion or a special cohesion failure at the track bridges and, in particular, at the left side of the first DAF. This failure is evidenced by parts of the sensor structure and track bridge being visible at the strap and the lap side. The latter failure mode effects the sensor integrity as well, and it is thus further discussed in Sect. 4.2.3.

The adhesion or special cohesion failures between the PEI-sensor and the PVDF cover occur in the second-generation smart inlay MDAFs with uncut PEI as well. Figure 4.19a depicts the entire bondline damage after the crack growth test. At both DAFs, the main failure mode is the adhesion or special cohesion failure of the PEI-sensor-PVDF interface. This is evidenced by most of the sensor structure being on the strap side and the rather smooth failure plane. In addition, an adhesion or a special cohesion failure of the PEI substrate and the sensor structures occurs at the start of both DAFs (see Fig. 4.19a and b) and at the track bridge (see Fig. 4.19a and c). This failure causes a separation of the PEI sensor so that parts of it are found on the lap and strap.

The described crack path differs from that of the other specimens. As illustrated in 4.19d, the crack still grows into the interface of the MDAF and the CFRP at first. However, it does not stay there but moves into the PEI-sensor-PVDF DAF interface

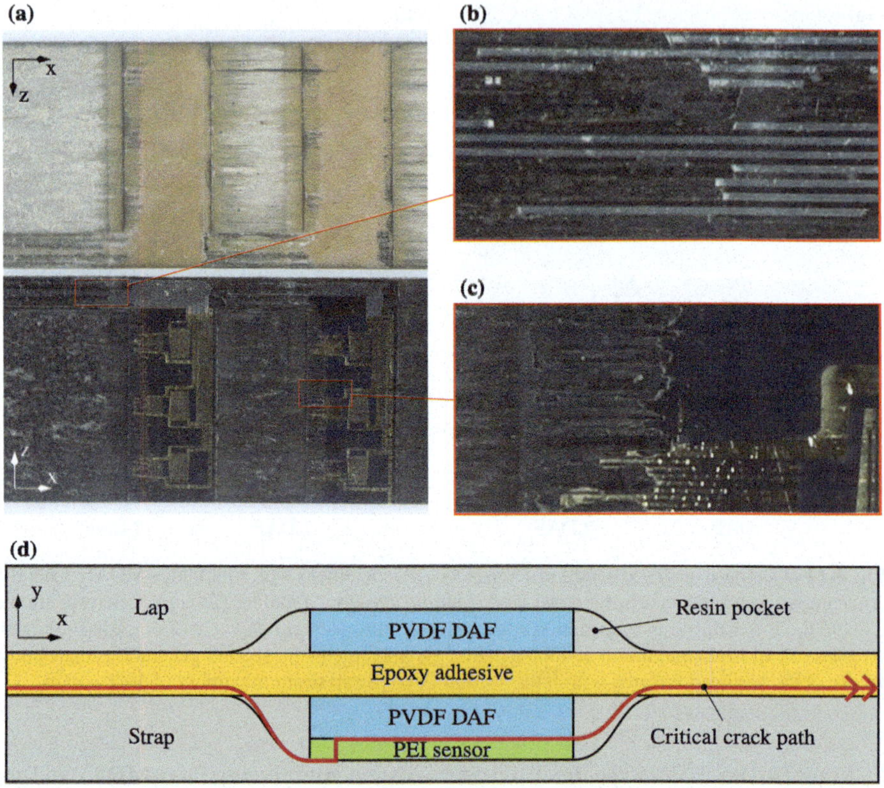

Fig. 4.19 Bondline damage in a specimen with 2nd-gen. smart inlay and uncut PEI after crack growth tests at 3000 $\mu m\,m^{-1}$ strap strain maximum **a** Specimen overview **b** Failures in the track bridge **c** Failed CFRP-PEI-sensor and PEI-sensor-PVDF interfaces **d** Crack path

of the MDAF after some distance. Remarkably, the crack does not immediately break the PEI-sensor-PVDF interface. A possible explanation for that is particularly high strains at the crack front in x- and y-direction. This is discussed further in Sect. 4.2.4, which deals with the sensor integrity.

The failure of the PEI DAFs occurs in the interface between the CFRP and the PEI. As depicted in Fig. 4.20, after the start of the first DAF, many first-ply fibres are pulled out of the strap. However, the damage does not grow deeper into the laminate but stays in the CFRP-PEI interface. Hence, the critical crack paths of the PEI and PVDF DAFs are the same.

The reason the CFRP-PVDF/-PEI and not the PVDF-/PEI-epoxy adhesive interface tends to fail is most likely due to the different surface pretreatments. While the thermoplastics in contact with the epoxy adhesive are plasma-treated before the bonding process, there is no plasma treatment before the co-curing process. Schollerer [Sch24, pp. 80–83] reports that plasma or Vacuum Ultraviolet (VUV) light treatments improve the adhesive strength in the co-cured CFRP-PVDF interface as well.

Fig. 4.20 Bondline damage in a PEI DAF specimen after cyclic loading at 3000 $\mu m\,m^{-1}$ strap strain maximum

PEI has not been investigated, but it seems plausible that the surface activation of PEI would result in even better adhesion, too. In conclusion, a surface activation before co-curing shows great potential to prevent the failures in the CFRP-PVDF/PEI interface. This in turn could keep the crack path in the epoxy adhesive and reduce the risk of failure in PVDF-based DAFs.

The crack path moving below the DAFs towards the strap noticed in all kinds of CLS specimens is relatively similar to the observations of Löbel [Löb16, pp. 139–141] in his CLS crack growths tests with hybrid bondline. Instead of a separation of the CFRP-PVDF interface, he observes delaminations below the strap for specimens tested at 4000 $\mu m\,m^{-1}$ and 5000 $\mu m\,m^{-1}$ strap strain maxima. Hence, the crack shows the tendency to grow towards the laminate in both cases. Delaminations occur in two out of three specimens tested at 4000 $\mu m\,m^{-1}$, whose crack in the bondline is arrested by the hybrid bondline DAF. Remarkably, no delaminations at the same strap strain maximum are observed in the nine specimens with surface toughening, whose crack is arrested in the first DAF. Why there are no delaminations in the specimen with PVDF surface toughening elements instead of hybrid bondline elements, even though the materials and manufacturing processes are identical, is not clear. One possibility might be that the hybrid bondline unloads the crack tip more than surface toughening. The consequence might be a higher load concentration in the laminate around the DAF, causing the delamination in the case of the hybrid bondline. There is no stress data to validate this assumption, but two observations of Löbel [Löb16, pp. 140–141] support it. The delamination is initiated by an interlaminar crack forming in the second and fifth plies above the welded PVDF area, but stops again at the end of the DAF, where there is no welded PVDF but only surface toughening [Löb16, Fig. 10.21]. If that turns out to be true, it could mean that there is an optimal stiffness of the DAF material at which the crack tip in the bondline is unloaded sufficiently for disbond arrest, and the stress concentrations in the surrounding laminate are low enough not to cause delamination. However, even if this assumption turns out to be true, the practical relevance of such a stiffness optimisation is in question because the load levels at which the delaminations occur are already at or above typical static ultimate strains of the laminate.

Altogether, the crack tip and bondline damage results of the CLS specimens with surface toughening elements show that the failure of the DAFs is caused by the adhesion or special cohesion failure near the CFRP adherend of the PVDF-based (M)DAFs. However, only in six out of fifty-six PVDF-based two-stage DAFs, the crack is only arrested in the second DAF or not at all. Considering that the specimens are tested at design limit strain levels and above, the disbond arrest functionality of the PVDF-based two-stage DAF can thus be confirmed for the tested configuration and manufacturing process. However, strengthening the CFRP-PVDF interface by a surface activation of the PVDF before co-curing is promising to further improve the performance of the PVDF-based (M)DAFs.

In contrast, the second-generation smart inlay MDAFs with uncut PEI do not show a disbond arrest function. The crack growth through the PEI-sensor-PVDF interface in all four specimens of this configuration indicates that the PEI-PVDF and sensor-PVDF interfaces are weaker than the PEI-epoxy and PVDF-epoxy interfaces and incapable of withstanding the cyclic loads for long. Considering that the crack is completely arrested in most specimens with second-generation smart inlay MDAF, whose PVDF is in contact with the laminate due to the cut PEI, supports that assumption. This also explains the deviations in the disbond arrest functionality of the second-generation smart inlay MDAFs because the long crack lengths are all obtained for the smart inlays with uncut PEI. However, since one time a disbond arrest function of two-stage MDAF is observed, even though at the end of the second DAF, it is open whether the cracks are slowed down in the other specimens with second-generation smart inlay MDAF at least.

Regarding the PEI DAFs, the explanation that a weak interface jeopardises the disbond arrest function is not plausible. The CFRP-PEI interface is reported to be particularly strong even without surface activation [Kyr+21, p. 17–18]. Hence, it is unlikely that the failure of the PEI DAFs is caused by an insufficient interface strength. Therefore, it is of interest to investigate further whether the PEI DAFs affect the crack growth at all.

Crack Growth in Specimens with Failed Disbond Arrest Feature

This section discusses whether the PEI DAF and second-generation smart inlay MDAFs, which do not show a disbond arrest function, affect the crack growth at

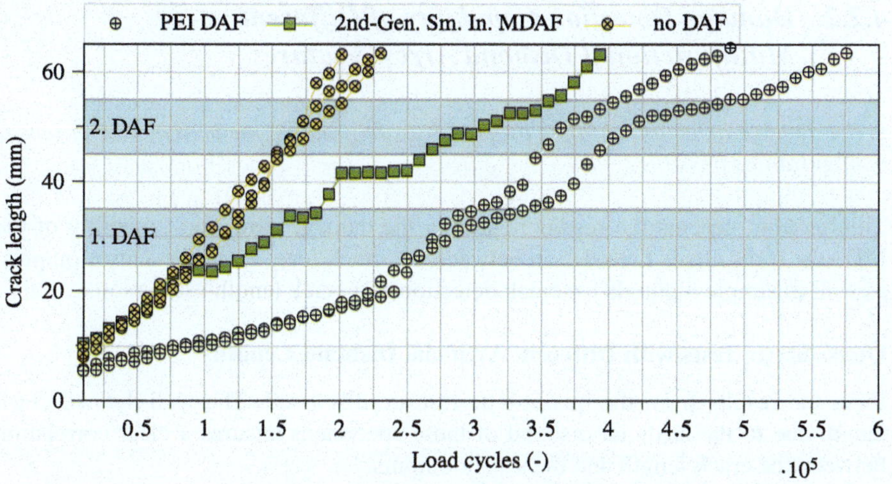

Fig. 4.21 Comparison of the crack growth of three specimens without DAF, two with PEI DAF, and one with 2nd-generation smart inlay MDAF and uncut PEI under cyclic loading at 3000 $\mu m\, m^{-1}$ strap strain maximum

least. For that matter, Fig. 4.21 plots the crack length of three specimens without DAF, two with PEI DAF and one with second-generation smart inlay MDAF and uncut PEI over the load cycles.

Even though the crack has not been arrested by the second-generation smart inlay MDAFs throughout the entire test, it has come to a temporary stop at both DAFs. Interestingly, a temporary stop is observed again at the end of the first DAF. Moreover, the temporary stop in front of the second DAF is indicated a few millimetres in front of the DAF. Since the crack length reading is only conducted from one side, the more likely reason is an underestimated crack length rather than the crack growth being stopped in front of the DAF already. Overall, it takes 175,000 cycles more for the crack to reach a length of 65 mm in the specimens with second-generation smart inlay MDAF compared to those without DAF.

In contrast, the PEI DAFs do not seem to slow down the crack growth. Both PEI DAF specimens show the slowest crack growth at first glance. However, this is due to the much more shallow crack growth rate in the areas with epoxy adhesive only. Within and near PEI DAFs the crack growth rate even increases instead of decreasing. Hence, the slower crack growth cannot be attributed to the presence of the PEI DAFs, but must be caused by a different effect. These results are in agreement with the claim of Schollerer [Sch24, p. 177] that PEI is too stiff for a surface toughening effect in structural bonds.

4.2.2 *Damage Detection Capability of Different Multifunctional Disbond Arrest Features*

Is the damage detection with the proposed strain gradient measurement and sensor configurations possible?

This section presents the results regarding the damage detection capability of the MDAFs. It discusses the relation between the crack length and the sensor outputs, as well as simple methods for crack detection and crack length estimation.

Quasi-Static Tests with Different Artificial Disbond Lengths

First, the results from the quasi-static test are discussed. The well-defined crack length due to the artificial disbond of these specimens ensures a clear correlation between the crack length and the sensor outputs.

Figure 4.22 illustrates the sensor signals of a CLS specimen with a 23 mm artificial disbond and a second-generation smart inlay MDAF. In addition to the output of each sensor, it shows the average of all sensors in the first and second DAF, respectively. At the 23 mm disbond length, the output of the first DAF sensors is significantly larger than that of the second DAF sensors upon loading.

Figure 4.23 shows bar plots of the averaged sensor in each DAF for 10 mm, 16 mm and 23 mm disbond lengths at the maximum load of 5.104 kN. The results for each artificial disbond length come from a single specimen. Since the load is applied three times for ten seconds, the error bars do not refer to multiple specimens, but to the

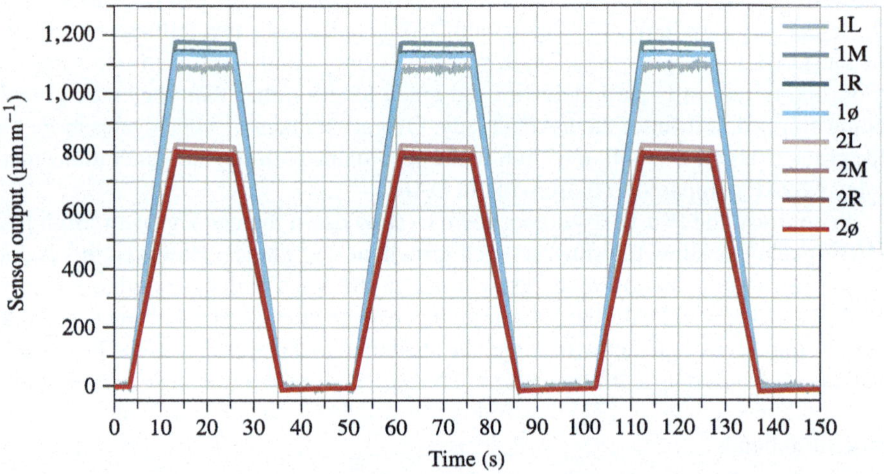

Fig. 4.22 Quasi-static loading of a CLS specimens with 2nd-generation smart inlay MDAF at an artificial disbond length of 23 mm. The labels of the sensors in the first and second rows are in accordance to Figs. 4.3 and 4.5. Averages are indicated by the ø-symbol

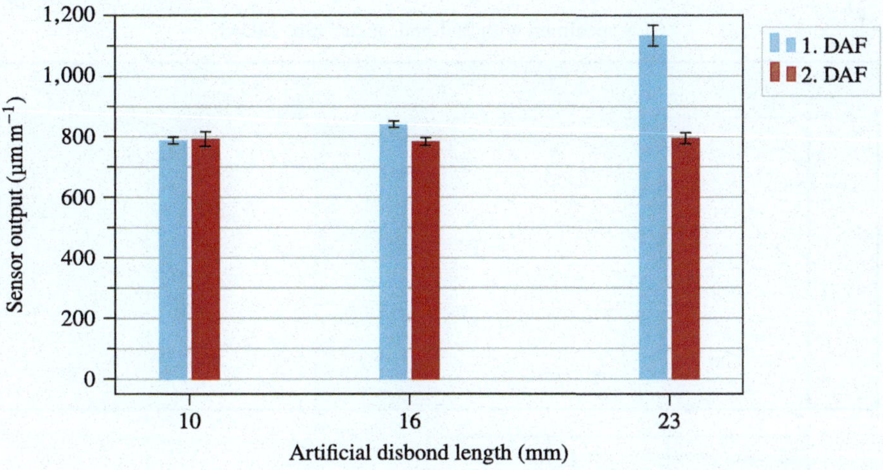

Fig. 4.23 Comparison of the averaged sensor outputs in the first and second DAF for specimens with second-generation smart inlay MDAF and three different artificial disbond lengths in the quasi-static test at 5.104 kN load

deviation of all three sensors per DAF over the thirty seconds the maximum load is applied in total.

The outputs of the first and second DAF sensors are the same for the disbond length of 10 mm. For the longer disbonds, the output of the first DAF sensors increases, whereas the second DAF sensors stay at the same level as before.

The presented measurement principle depends on the distance between the sensors and the damage. This correlates with the artificial disbond length or crack length,[24] but for a more intuitive understanding, the artificial disbond distance (crack distance) is introduced in Fig. 4.24 for the evaluation. For the sensor position, the centre of the measurement grid in the x-direction is chosen. This accounts for the fact that the measurement grids integrate the strain over their length, and that the positioning of the sensors changes slightly between the MDAF variants. Measured from the overlap edge of the CLS specimen, the sensor positions are approximately 27 mm for the first and 47 mm for the second DAF in the case of the depicted second-generation smart inlay MDAF. For the FBG MDAF, however, the distances are 30 mm and 50 mm, and for the third-generation smart inlays 29 mm and 49 mm, respectively. The artificial disbond distance (crack distance) is the difference between the sensor position and the corresponding artificial disbond length (crack length). With that definition, the results can be evaluated.

The results confirm the assumption that the strain distribution in the MDAF is a valid indicator for detecting disbond in its vicinity. They are in good agreement with the simulation of the longitudinal strain in CLS specimens with hybrid bondline and

[24] This distinction accounts for the fact that the damage is an artificial disbond in the CFRP-epoxy interface in the quasi-static test, but a crack in the bondline during the crack growth test.

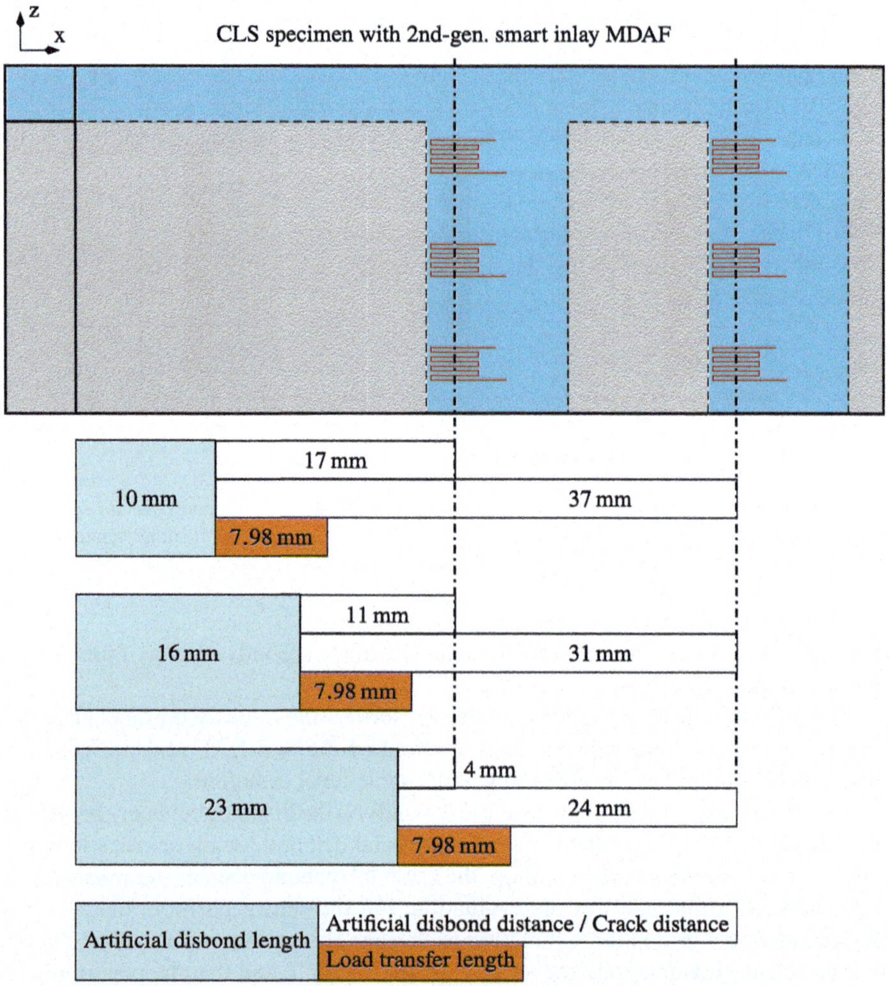

Fig. 4.24 Distance between the sensor grids of the smart inlay MDAF second generation and the artificial disbond length in correlation with the load transfer length (see Eq. 4.4)

similar artificial disbond lengths (see Fig. 3.6), and the theory regarding the load transfer length presented in Sect. 4.1.2. The latter explains well, why the sensor outputs are similar at an artificial disbond length of 10 mm but start diverging at the larger artificial disbond lengths. The artificial disbond distances at the 10 mm artificial disbond length, are approximately 17 mm and 37 mm, respectively. As illustrated in Fig. 4.24, both distances are far greater than the estimated load transfer length in the CLS specimen of 7.98 mm (see Eq. 4.4). At the artificial disbond lengths of 16 mm and 23 mm the corresponding distances regarding the second DAF sensors are 31 mm and 24 mm, respectively. This is still much greater than the estimated load transfer length itself. In contrast, the artificial disbond distance regarding the first DAF sensors drops to 11 mm for an artificial disbond length of 16 mm. In this condition, the load transfer seems incomplete, even though the artificial disbond distance is still slightly larger than the estimated load transfer length. Accordingly, the output of the sensors in the first DAF shows a small increase. At the artificial disbond length of 23 mm the artificial disbond distance in the first DAF drops to 4 mm, and thus below the load transfer length. This means the load transfer is definitely not completed before the sensors in the first DAF are reached. As a result, the output rises significantly. Since the sensors in the second DAF do not change, the signals diverge clearly.

Crack Growth Tests

The following discusses whether the findings from the quasi-static tests with defined artificial disbond lengths are reproducible in the crack growth test. The crack growth tests provide more realistic crack fronts, which is why they are more representative for the evaluation of the damage detection capability.

Figure 4.25 shows the sensor output of a CLS specimen with FBG MDAF and the crack length throughout a complete run of the crack growth test. The maximum strap strain level of this specimen is $3000\,\mu\text{m}\,\text{m}^{-1}$, and the plotted sensor output is recorded at mean load, which equals a strap strain of $1650\,\mu\text{m}\,\text{m}^{-1}$. For better visibility, only every 25th measurement point is included in this overview. Figure 4.26 depicts a detailed view of the same plot, containing the first 90,000 load cycles and including all measurement points.

The outputs of the sensors placed in the first and second DAF start at approximately half the strain introduced into the strap. The detailed view shows a slight difference between both signals, with the sensor in the first DAF returning a higher value. The

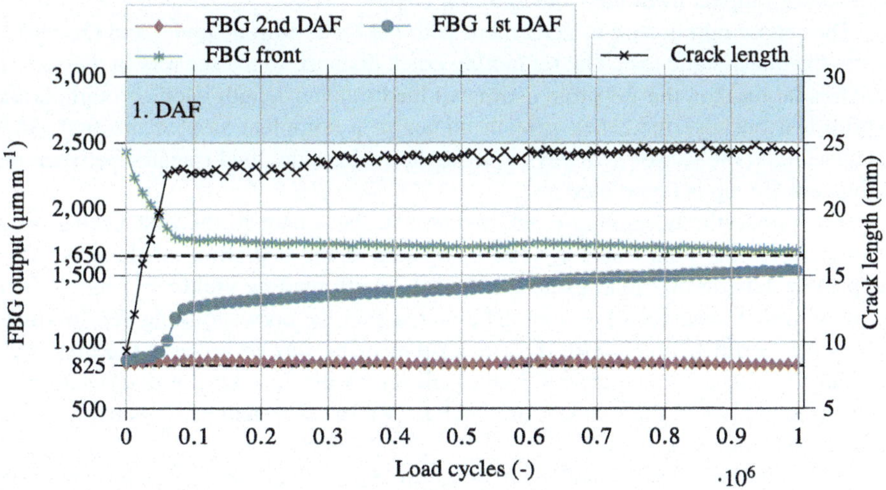

Fig. 4.25 Complete crack growth measurement recorded at mean load of a specimen with FBG MDAF tested at $3000\,\mu\text{m}\,\text{m}^{-1}$ strap strain maximum level

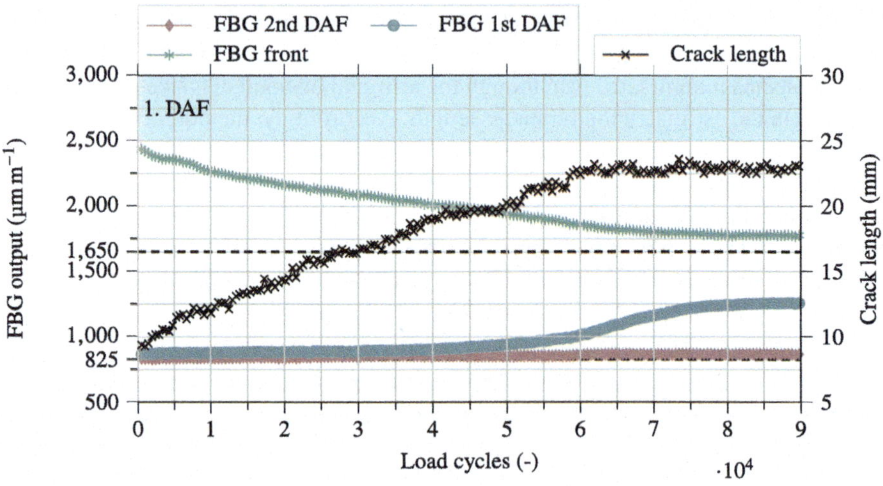

Fig. 4.26 Detailed view showing the first 90,000 cycles of Fig. 4.25

sensor placed in the front, directly at the artificial disbond, returns a strain value of $\approx$2500 µm m^{-1}, which is even much higher than the strap strain of 1650 µm m^{-1} itself. When the crack grows towards the first DAF, the signals of the sensors in the front and the first DAF converge towards the strap strain level. However, the sensor in the first DAF does not reach it completely until the end of the test. In contrast, the output of the sensor in the second DAF remains at its initial value throughout the entire test. The crack appears to progress very slowly but continuously after its quick growth within the first 70,000 cycles. This correlates with the continuous changes of the sensor outputs from the front and first DAF.

The sensor outputs are in agreement with the load transfer model and the results from the quasi-static test. The remaining crack distance to the sensors in the second DAF is larger than the 7.98 mm estimated load transfer length itself throughout the entire test because the crack growth comes to a complete stop at the first DAF. Accordingly, the sensor is always at a location where the load transfer between the strap and the lap is completed.

In contrast, the remaining crack distance to the sensor in the first DAF is only sufficient for the load transfer to be completed at the beginning of the test. After approximately 40,000 load cycles, the crack length reading yields $\approx$18 mm, which equals a crack distance of 12 mm. The increase in the corresponding sensor signal thereafter correlates well with the crack tip growing further towards the first DAF. At this stage, every crack growth reduces the crack distance until it reaches approximately 6 mm when the crack is arrested at 24 mm crack length. This explains the increased output signal well.

However, the sensor signal seems to saturate later than the crack length reading. This indicates that the crack front along the specimen width might not be straight, but further ahead at the edges of the specimen. Löbel [Löb16, p. 134] reports of

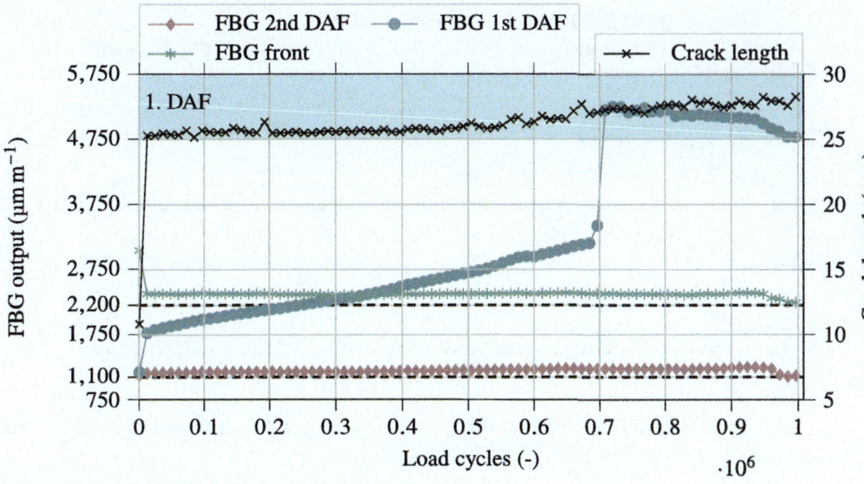

Fig. 4.27 Complete crack growth measurement recorded at mean load of a specimen with FBG MDAF tested at 4000 μm m^{-1} strap strain maximum level

an angular or even convex crackfront, whose crack is further ahead in the centre instead of the edges, in similar experiments. The different observations show that a measurement on the side of the specimen is insufficient to accurately determine the crack length in the centre.

The fact that the sensor in the first DAF does not reach the strap strain value until the end of the test is also explainable by the load transfer model. At the end of the test, the crack length reading is approximately 25 mm, which yields a crack distance of 5 mm. Hence, there is still a sufficient distance for some load transfer. Accordingly the strain at the sensor is less than the strap strain.

Regarding the sensor in the front of the specimen, the decrease in its output also correlates with the crack growth. In this case, it is assumed that strain peaks around the crack tip cause the extremely high output of this sensor at the beginning of the test. As the crack tip leaves the sensor area, the signal decreases toward the expected value of the strap strain, which leads to the observation that the signals from the sensors in the front and in the first DAF converge.

Figure 4.25 also illustrates the potential of the FBG sensors to withstand the cyclic loads at a maximum strap strain of 3000 μm m^{-1}. The sensors provide realistic strain outputs with integrated FBG sensors until the end of the crack growth test. Section 4.2.3 further discusses the sensor integrity.

Figure 4.27 depicts the results of the crack growth measurement of another specimen with FBG MDAF. In contrast to the previous specimen, the maximum and mean strap strains are 4000 μm m^{-1} and 2200 μm m^{-1}, respectively. In principle, the correlation between the crack length and the sensor outputs is similar to the one obtained for the previous specimen, as the detailed view in Fig. 4.28 shows. However, in this specimen, the crack reaches the first DAF much sooner, and it starts penetrating it.

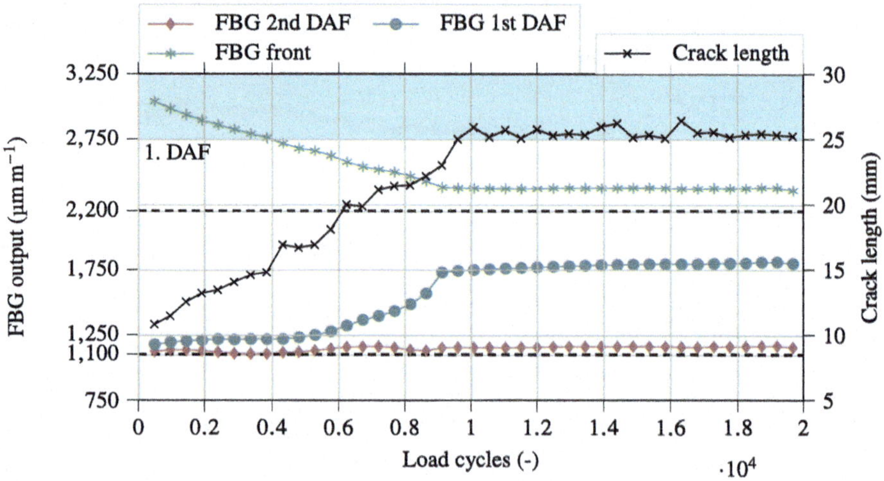

Fig. 4.28 Detailed view showing the first 20,000 load cycles of Fig. 4.27

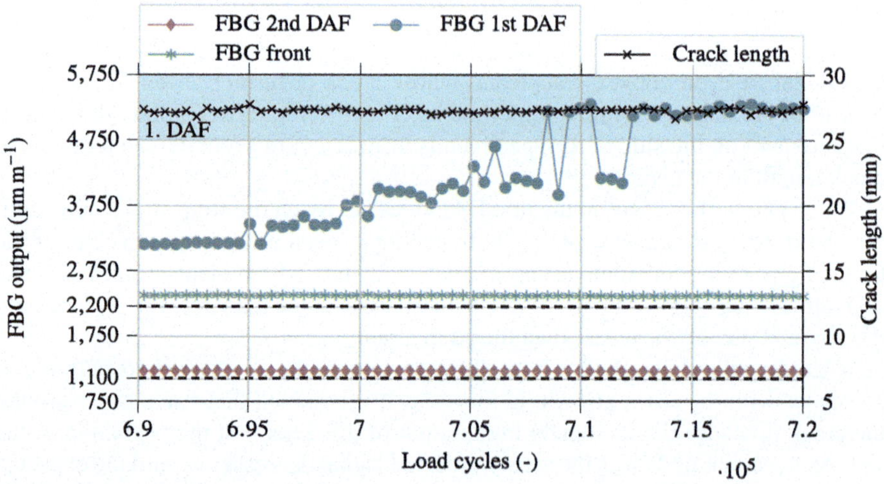

Fig. 4.29 Detailed view showing the load cycles 690,000 to 720,000 of Fig. 4.27

Hence, the sensors in the front and the first DAF change much faster, too. In addition, the sensor in the first DAF does not saturate near the mean strap strain level but grows at a constant rate until it suddenly increases to a value above 5000 μm m^{-1}. Moreover, all FBG sensors drop simultaneously shortly before the end of the test.

The different behaviour of the sensor in the first DAF correlates with the crack growth, which does not stop once the first DAF is reached, but slows down significantly. The rapid increase in the sensor output to over 5000 μm m^{-1} occurs shortly after the crack length reading on the side of the specimen has reached approximately

27 mm. As the detailed view in Fig. 4.29 demonstrates, the sensor signal does not jump as it appears in Fig. 4.27. Moreover, the crack length reading on the side of the specimen does not change according to the detailed view when the sensor output starts rising. As discussed before, the crack growth in the centre is probably a little behind. In addition, the crack length algorithm is not completely accurate (see Sect. 4.1.4). The most plausible explanation for the major increase in the sensor output is, thus, that the crack length in the centre of the specimen reaches the measurement grating of the FBG sensor, which starts 28.5 mm away from the overlap edge, at this point. This is followed by the same increase in the strain as observed for the sensor in the front. The fact that the sensor in the front is placed on the stiffer epoxy resin instead of the PVDF material explains why the output of the sensor in the first DAF exceeds the maximum of the sensor in the front.

There is no correlation between the simultaneous drop of all FBG sensors at the end of the test and the crack growth. The simultaneous drop in all three sensor signals indicates a systematic influence on the entire measurement system. Hence, it is not plausible that the changes correlate with further crack growth. Moreover, the force and displacement output of the test rig do not show anomalies and are thus also ruled out as a cause. Due to the long time of the measurement, a temperature change is a plausible explanation, but there is no evidence for that as there is no temperature measurement. Since the outputs do not change to unrealistic values, it can be concluded that the FBG sensors are also capable of withstanding the higher loads at the increased strap strain maximum level of 4000 $\mu m\, m^{-1}$.

Figures 4.30 and 4.31 illustrate the crack growth measurement and a detailed view of it for a specimen with a third-generation smart inlay MDAF. This particular specimen has no surface toughening element in the lap. Neither specimen with a

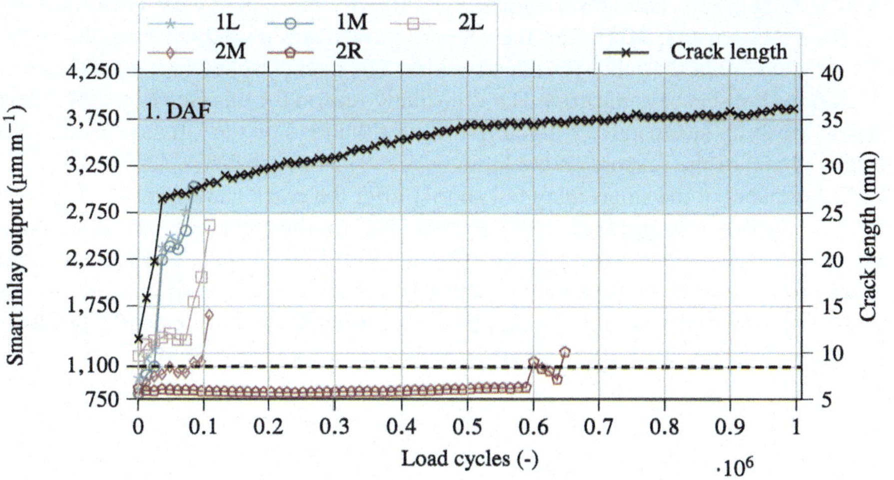

Fig. 4.30 Complete crack growth measurement recorded at mean load of a specimen with 3rd-generation smart inlay MDAF tested at 4000 $\mu m\, m^{-1}$ strap strain maximum level

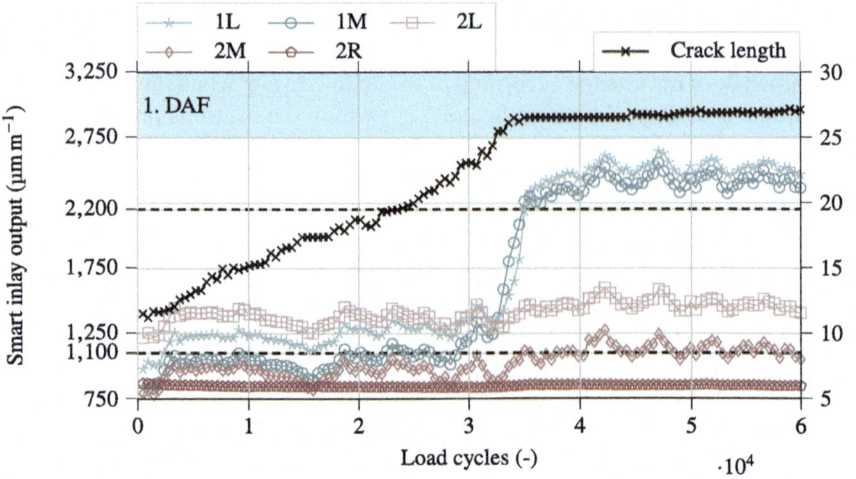

Fig. 4.31 Detailed view showing the first 60,000 load cycles of Fig. 4.30

smart inlay MDAF consists of a sensor at the artificial disbond. Instead, there are three sensors in the first and second DAF each (see Fig. 4.5). Moreover, the sensor number 1R of the illustrated specimen has been damaged during manufacture and is thus not depicted. The maximum and mean strap strains of this specimen are 4000 μm m^{-1} and 2200 μm m^{-1}, respectively.

The correlation between the sensor output of the smart inlay and the crack length is similar to the FBG measurements. Again, the sensors in the second DAF remain mostly constant, whereas the sensors in the first one show a rising signal at an approaching crack. The sensor signals vary slightly more than what been observed for the FBG sensors. Moreover, the sensors show a strain gradient from the left to the right side regarding the specimen's width. The pocket milled into the left side of the lap for the sensor connection is a very likely reason for this (see Fig. 4.6b). As a result, the strap has to carry a higher portion of the load on the left side. Hence, the higher signal of the sensors on the left.

The sensors of the smart inlay fail shortly after the crack has approached the first DAF. Only the sensor placed in the second DAF on the right side outputs realistic strains after the crack has reached the DAF, but it still fails long before the end of the crack growth test. For that reason, it cannot be observed how the signals change due to the crack growing through the first DAF. See Sect. 4.2.3 for a detailed analysis of the sensor failures.

Damage Detection Approaches

The experimental results demonstrate the sensitivity of the integrated sensors to damage in the bondline in dependence on the sensor positions. This section further discusses approaches to use this effect for damage detection and crack length estimations.

A straightforward approach to assess the divergence of the sensor outputs is utilising the signal differences as a damage criterion. Figure 4.32 expands Fig. 4.23 by the differences between the averaged sensor outputs. For the three tested artificial disbond lengths, the difference is nearly zero for the shortest one at 10 mm, increases slightly at 16 mm, and becomes clearly noticeable at 23 mm when the disbond is only 2 mm away from the first DAF.

Figure 4.33 illustrates that the difference in the sensor outputs depends linearly on the applied load and not only the crack length. It shows the differences between the averaged first and second DAF sensors during a loading ramp of the quasi-static test at all three artificial disbond lengths. With increasing artificial disbond length, the gradient of the output difference increases noticeably. For the shortest artificial disbond, it is nearly zero, whereas, for the longest one, it rises above 300 μm m^{-1}.

The results from the quasi-static tests show a significant change in the difference of the sensor outputs due to the changed artificial disbond length. However, the difference also depends on the applied load, and the quasi-static measurement is conducted at well-defined damages. The crack growth data thus provides information on the feasibility of utilising the output difference for more realistic loading conditions and damage sizes.

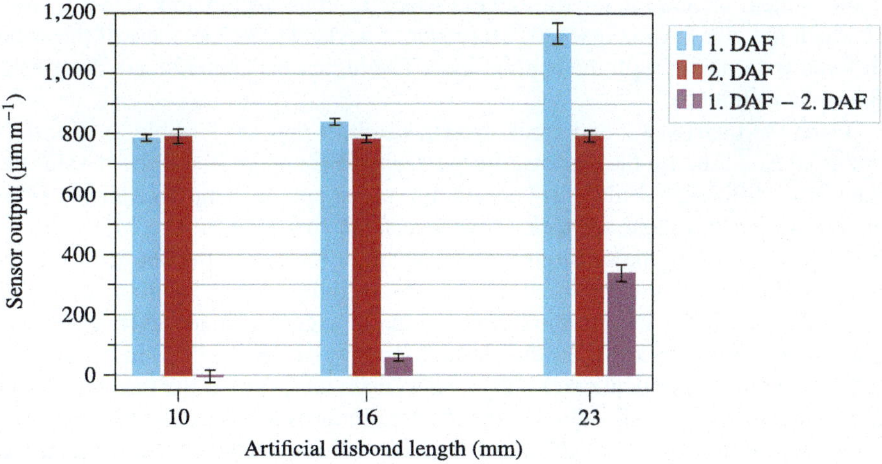

Fig. 4.32 Difference of the averaged sensors in the first and second DAF from Fig. 4.23 for three different artificial crack lengths at 5.104 kN load (mean load for the strap strain maximum of 3000 μm m^{-1})

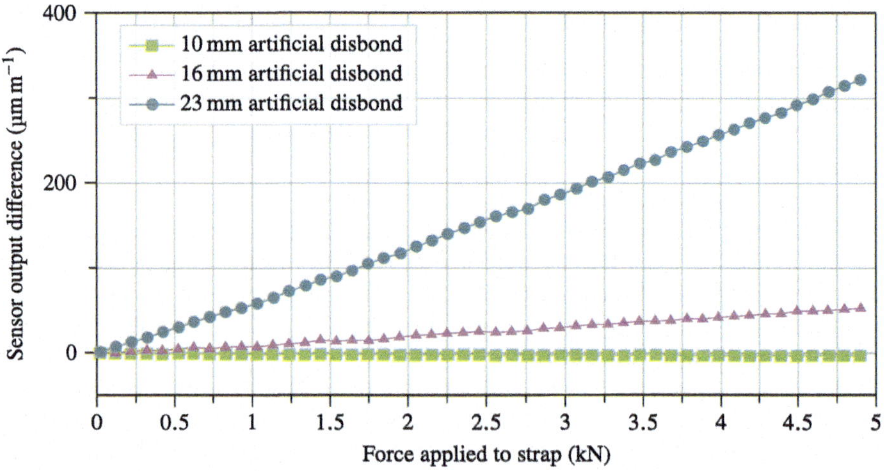

Fig. 4.33 Averaged second DAF sensors subtracted from the averaged first DAF sensors during a loading ramp from 0 to 5 kN for three different artificial disbond lengths in the quasi-static test

The findings from the quasi-static measurement are reproducible in the crack growth measurements. Figure 4.34 plots the differences between the first and second DAF FBG sensors from Fig. 4.27 over the crack length for the minimum, maximum, and mean load applied to the specimen during cyclic loading at 4000 $\mu m\,m^{-1}$ strap strain maximum. Due to the noise in the crack length readings in particular, multiple strain differences can be assigned to the same crack length. However, the plot still shows a clear increase in the differences when the crack length approaches 25 mm, at which the first DAF is reached. Moreover, it is visible that the value of the strain difference correlates with the applied load. The higher the applied load, the greater the strain difference.

From the quasi-static and crack growth result, it appears feasible to identify a damaged bondline by the increase in the sensor difference. A threshold value for the sensor difference could thus be used as a rather simple damage criterion. Upon exceeding the threshold, a crack growth near the first DAF of the MDAF could be indicated. Due to the sensitivity of the strain difference to the applied load, the threshold has to be defined with care. The load sensitivity has two main consequences: At high loads, an increase in the difference might trigger an alarm even though the crack length is not critical if the threshold is too low. At low loads, an increase in the difference due to a critical crack size could remain unnoticed if the threshold is too high. As indicated in Fig. 4.34, for that CLS specimen, a threshold of 200 $\mu m\,m^{-1}$ seems appropriate. At 12.43 kN, this threshold would be reached at a crack length of approximately 17 mm, and at 1.243 kN, it would be reached at approximately 22 mm crack length.

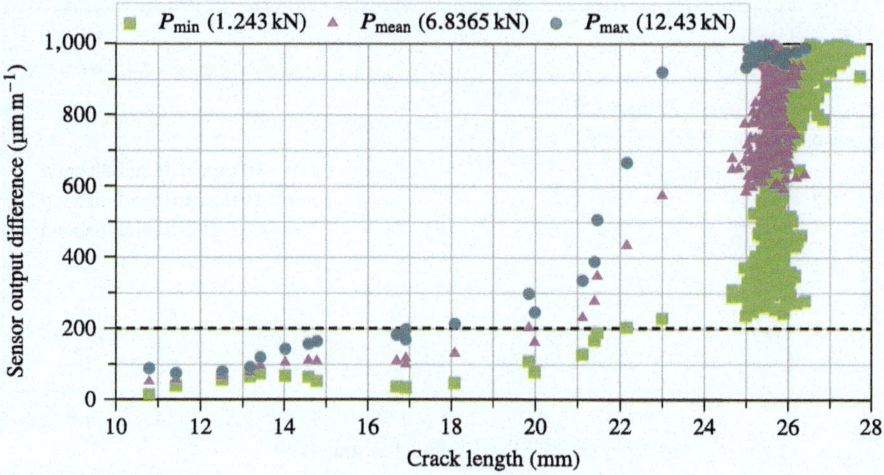

Fig. 4.34 Difference of the FBG sensor outputs over the crack length during cyclic loading of a single specimen at the min, mean, and maximum load regarding the strap strain maximum of 4000 μm m^{-1}. The output of the sensor in the second DAF is subtracted from the output of the sensor in the first DAF. Possible damage threshold marked by the dashed line

Von der Heide et al. [Hei+22, p. 13–15] propose a different evaluation method to reduce the sensitivity to the applied load. It uses the same sensor outputs, but it relies on evaluating the quotient of the signals instead of the difference. The idea is that if both strains are caused by the same load, the load cancels out when taking the signal quotient. As a result, the quotient would depend on the crack length only.

Figure 4.35 supports the feasibility of that idea. It shows the signal of the averaged first DAF sensors divided by the averaged second DAF sensors during a loading ramp in the quasi-static test. It uses the same data as Fig. 4.33. In contrast to the signal difference, the signal quotient does not show a clear dependency on the applied load. For the loads near zero, there is a significant change in the quotient, though, and there is a slight increase in the quotients at 16 mm and 23 mm artificial disbond. The reason for the increase at low loads is very likely that the quotient has a singularity at zero strain, or rather zero voltage across the Wheatstone Bridge of the strain gauges. Hence, there is the risk that the quotient shows strong deviations if the sensor outputs are low.

This effect is clearly visible in the snapshot measurements during cyclic loading. As indicated in Fig. 4.36, the quotient during cyclic loading is not constant but increases noticeably when the load drops towards 1 kN. In this case, it rises from a value of about two to over eight. A value of two is realistic when the crack has reached the DAF because the load on the sensor in the first DAF is approximately twice as big then. Hence, the quotient criterion appears to be risky in terms of false alarms, if the quotient becomes large at low loads and small or no damage.

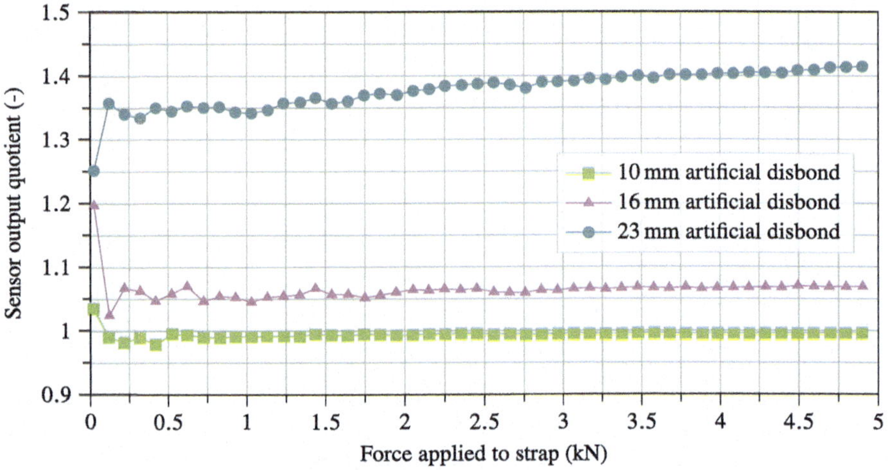

Fig. 4.35 Averaged first DAF sensors divided by the averaged second DAF sensors during a loading ramp from 0 to 5 kN for three different artificial disbond lengths in the quasi-static test

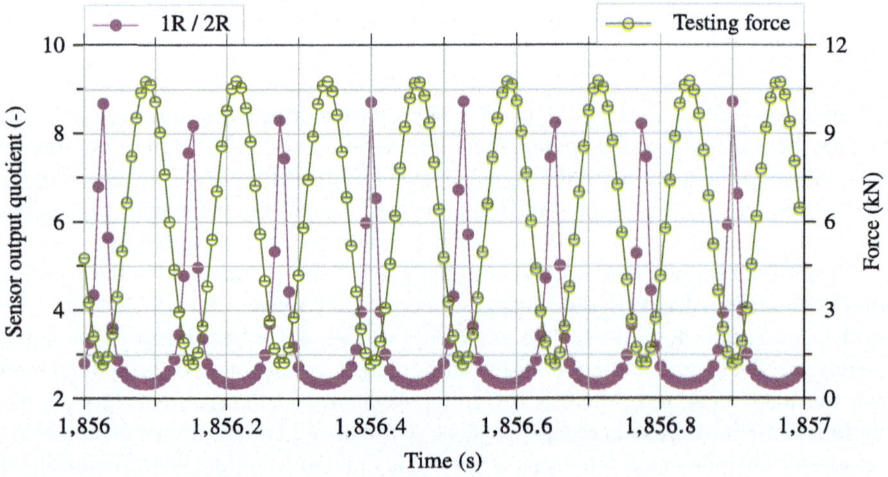

Fig. 4.36 One second of the signal quotient and applied force during sinusoidal loading of a specimen at maximum strap strain level of 3500 $\mu m\,m^{-1}$ and a crack length reading of 21.7 mm

Von der Heide et al. [Hei+22, p. 13] show the increase in the signal quotient at low loads as well. Taking the more stable values at high loads, they present an algorithm to utilise the signal quotient for determining the remaining distance of the damage to the first DAF [Hei+22, p. 14–15]. Information regarding the remaining distance seems attractive. For a robust algorithm that functions truly independently of the applied load, the challenge regarding the singularity of the quotient criterion at low sensor outputs has to be solved. One possibility might be using the strain information for estimating the applied load. This would mean the quotient would only be evaluated if both sensors exceed a defined strain value. Moreover, a method to deal with the effect of secondary loads that are not equal at both sensor positions has to be found, as they could cause large disturbances of the quotient.

The experimental results indicate the feasibility of damage detection in an overlap joint with a simple strain gradient approach in the strap. However, the result is obtained in a specimen with an unambiguous crack growth path. Hence, for applications with more complex load cases and undefined crack growth paths, more information on the load transfer might be necessary.

One option is utilising sensors in the load-receiving part of the overlap joint as well. In the case of the CLS specimen, this means evaluating the strain in the lap. Figure 4.37 illustrates the advantage of that method. It shows the VSG outputs from the DIC measurement for a specimen, whose crack passes both DAFs in the crack growth test. To reduce the scatter of the VSGs, a moving average of ten is used. Moreover, only every tenth data point is plotted.

The strains in the strap are similar to the previously discussed results, but there are some important differences. As before, the strain values approach the strain introduced into the strap with increasing crack length. In contrast to the sensors in the bondline, there are no elevated strains near the crack tip. Moreover, there is a much larger offset between the sensors at low crack lengths. Both effects can be explained by the fact that the strains are not evaluated in the bondline but in the middle of the CFRP adherends. There is no strain peak in the much stiffer CFRP due to the crack tip in the bondline, and the bending of the CLS specimen caused by the eccentric tensile load has a larger effect because the sensors are further away from the neutral axis.

The strains in the lap show the opposite behaviour. They start at approximately half the strap strain and reduce below zero when the crack passes their position. This is because the lap is unloaded above the crack. The negative strains originate from the compression of the sensors due to the bending load. The influence of the bending on the sensor outputs is illustrated in Appendix B.

These results indicate some limitations in measuring the strain in the strap only. If the crack passes both DAFs, the outputs of the sensors in both DAFs become similar again. If the load is unknown, neither the difference nor the quotient would be distinguishable from the undamaged state. Moreover, the bending load causes a strain gradient at the beginning of the crack growth test when the crack length is only 10 mm. Since bending loads cause strain gradients in undamaged specimens, too, using the damage criterion of a simple strain gradient measured with two sensors will not suffice for reliable damage detection if there are similar strain gradients due to a bent healthy and an elongated damaged structure. In the CLS configuration, however, damage detection is possible despite the superimposed strain gradients due to the bending load. Figure 4.37 illustrates that the original difference in the strain signals is approximately $200\,\mu\text{m}\,\text{m}^{-1}$, but when the crack reaches the first DAF it is approximately $600\,\mu\text{m}\,\text{m}^{-1}$. Hence, it seems feasible to identify the damaged bondline section with the proposed difference criterion, but the threshold would have to be increased to compensate for the strain gradient of the bending load.

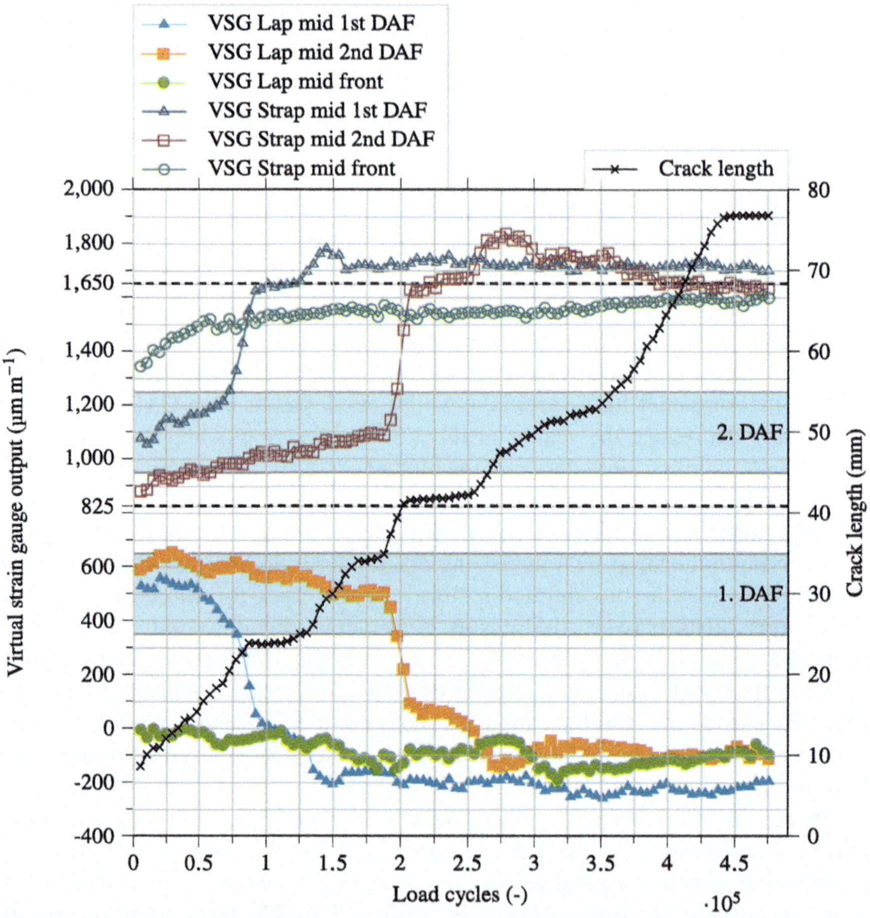

Fig. 4.37 Change of VSG outputs in the lap and strap during the crack growth test of a specimen containing a second-generation smart inlay with uncut PEI

The additional strain information from the lap solves the issue regarding the similar strain differences in the strap for no crack and a crack that has passed both sensors. As visible in Fig. 4.37, the strains in the strap and lap diverge due to the separation of the parts. Hence, it becomes clear if damage in the bondline prevents the load transfer from the lap to the strap or not.

The additional sensors, however, would increase the number of interfaces, and thus create possible weak points. Hence, a careful trade-off between the number of implemented sensors is necessary to meet the requirements for robust damage identification and minimal disturbance of the bondline. In the case of the CLS specimen with the sensors placed close to the neutral axis, the tensional load dominates the bending load even more so that damage is still detectable by two sensors in the strap despite the superimposed bending load. Moreover, the crack growth is stopped long enough to ensure damage detection before the crack passes the entire MDAF. In this particular scenario, two sensors in the strap seem to be a good trade-off. Section 5.1 discusses alternative ways to increase the information than adding more sensors to the bondline.

4.2.3 Sensor Integrity under Cyclic Loading

Are the integrated sensors capable of withstanding the mechanical loads in the bondline?

The previous section demonstrates that the sensors are capable of damage detection in CLS specimens. However, the smart inlays in particular do not function until the end of the crack growth test. It has to be emphasised, though, that applying the limit strain repeatedly at this frequency are extreme conditions. Regarding the FBG sensors, the results from the previous section already indicate their capacity to withstand the cyclic strains. Nevertheless, in other specimens, some more severe deviations of the FBG sensors from the expected values are observed as well. Hence, this section further discusses the sensor integrity throughout cyclic loading.

Figure 4.38 shows a severe deviation of a single measurement grating in an FBG MDAF specimen. Here, the output from the sensor in the first DAF shows a large decrease starting after approximately 400,000 load cycles. As illustrated in Fig. 4.39b, the first DAF of this specimen fails in the CFRP-PVDF interface. Figure 4.39d further depicts that the FBG fibre is attached to the lap side after the test. Since the sensor

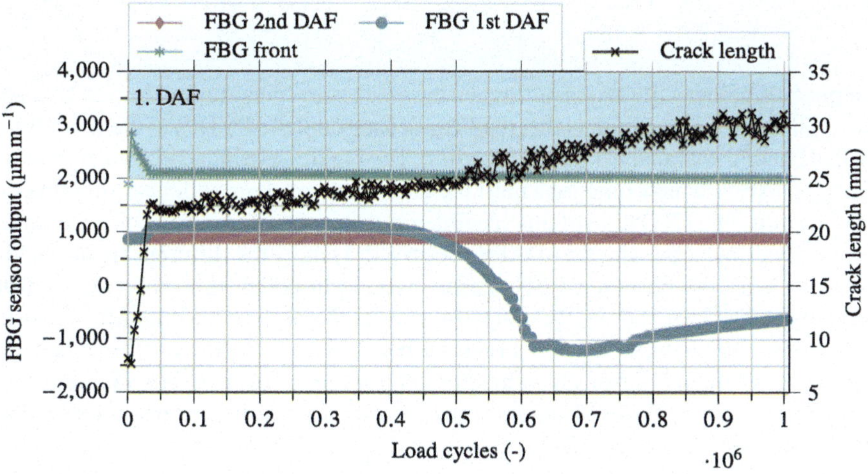

Fig. 4.38 Sensor failure of a single measurement grating in an FBG MDAF

Fig. 4.39 Comparison of the bondline damage in CLS specimens with FBG MDAF after the crack growth test and consecutive quasi-static loading until global failure **a** Specimen, whose FBG sensor stays in the strap **b** Specimen, whose FBG sensor is pulled out **c** FBG in the strap **d** FBG sensor in the lap

output falls below $-1000\,\mu\text{m}\,\text{m}^{-1}$ when the crack starts penetrating the first DAF, the most plausible explanation is that the FBG sensor (and the PVDF) at the first DAF is pulled out of the strap during cyclic loading. The bondline damage in a specimen with FBG MDAF[25], which functions until the end of the crack growth test, supports this explanation. The FBG fibre of this specimen is still in the strap, and a cohesion failure of the bondline above the first DAF is observed. Hence, the different failure modes correlate with the observed FBG sensor outputs. In conclusion, the crack growth tests of the FBG MDAFs suggest that the sensors themselves have a sufficient strain capacity for the integration into structural bonds. However, there is the risk of the sensor being pulled out of the adherend upon a failure of a DAF, which threatens their capability to provide strain information after a crack has grown into or passed a DAF with an embedded FBG sensor.

[25] This is the specimen analysed in the previous section (Fig. 4.27.)

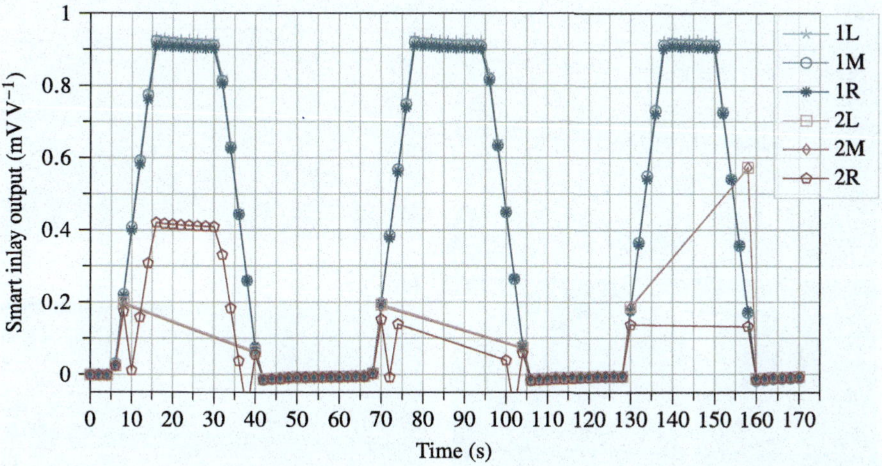

Fig. 4.40 Sensor failures in a quasi-static test of a specimen with first-generation smart inlay MDAF and 23 mm artificial disbond length

This strengthens the argumentation that a cohesion failure of the epoxy adhesive is the preferable failure mode. If different failure modes occur, it has to be ensured that the sensors provide sufficient information for identifying a damaged section of the structural bond. The possibility that the damage path might be in the bondline but below the sensor has to be considered for a reliable SHM system of the bondline.

The strain capacity of the third-generation smart inlay is noticeably lower compared to the FBG sensors. As already indicated for a specimen with third-generation smart inlay MDAF in Sect. 4.2.2, the sensors of the smart inlay show failures shortly after the crack has reached the first DAF. The sensor integrity of all three smart inlay generations is, thus, discussed in more detail to explain this behaviour.

The first-generation smart inlays show the worst integrity of all integrated sensors. Sensor failures occur already during the settling ramps before the cyclic loading or after a few load cycles. Figure 4.40 illustrates the sensor failures in a specimen with 23 mm artificial disbond in a quasi-static test. While the sensors in the first DAF stay intact throughout the entire test, all sensors in the second DAF fail quickly.

The most likely cause of the observed failure is damage to the track bridge. Figure 4.41 depicts the bondline damage in the same specimen after the crack growth test and quasi-static loading to global failure. Figures 4.41a and b illustrate severe damage to the track bridge. In contrast, there is no clearly visible damage to the measurement grids as they have been pulled out of the strap together with the PVDF substrate (see Fig. 4.41c). As a result, the track bridges are the most likely cause of the early failure in the quasi-static test before cyclic loading. The fact that the signal of the sensors in the second DAF is lost first is also in agreement with this argumentation. If the cause were failing sensor structures, the sensors in the first DAF should fail first, as they are subjected to a higher strain when the crack approaches. Failure of the track

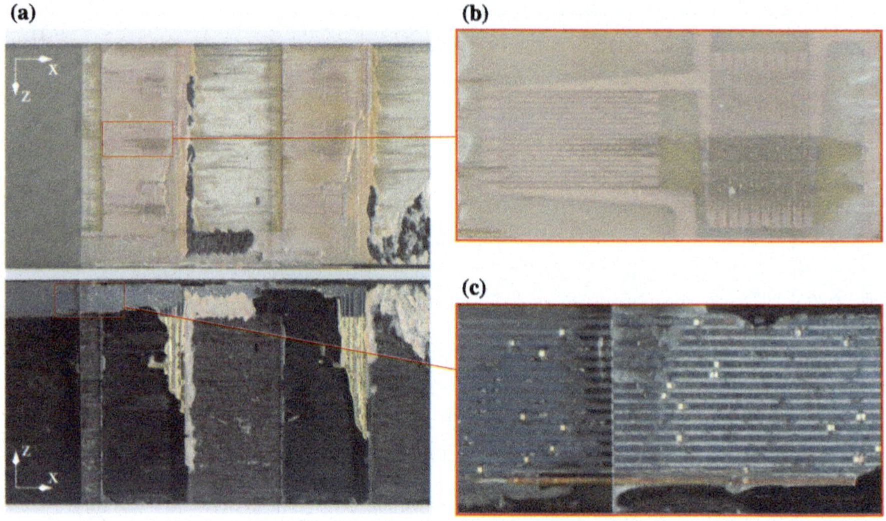

Fig. 4.41 Bondline damage in a specimen with first-generation smart inlay MDAF after a crack growth test at 4000 $\mu m\,m^{-1}$ and consecutive quasi-static loading until global failure **a** Overview **b** Optically intact sensor grids **c** Failure of the track bridge

bridge thus provides a plausible explanation as to why the sensors in the second DAF are the first to fail.

The second-generation smart inlays withstand more load cycles but suffer from early failures in the track bridge as well. Figure 4.42 illustrates the sensor outputs and crack length reading for a second-generation smart inlay with uncut PEI during the first 100,000 load cycles of a crack growth test. In contrast to the first-generation smart inlays, output signals are obtained for approximately 20,000 load cycles before sensor failures occur. However, the sensors still fail long before the crack reaches the vicinity of the first DAF. Accordingly, no divergence is visible between the recorded sensor signals of the first and second DAF. As a result, this generation does not allow for damage detection either.

The damage pattern of the specimen supports the interpretation of failed track bridges. Figure 4.43 depicts the damage in the specimens after the crack growth test. Figure 4.43a shows severe damage to the track bridge. In addition, there is severe damage to the sensor grids themselves (see Fig. 4.43b). The damaged track bridge, however, lies in-between the artificial disbond and the first DAF and is thus expected to fail first.

The results from the third-generation smart inlay MDAFs confirm the early failures of the track bridge as the main cause. Moving the track bridge away from the crack growth zone to the back of the CLS specimens significantly improves the sensor integrity under cyclic loading. As a result, the sensors of the smart inlays function until the crack reaches the first DAF and are capable of detecting that event. However, the sensors of the smart inlays seem incapable of withstanding the high strains near the crack front for long. Figure 4.43c illustrates characteristic damage to the sensor

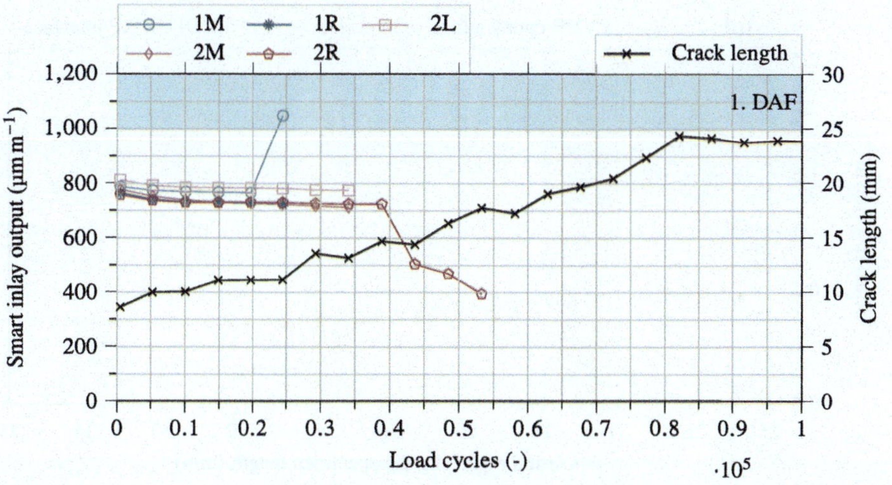

Fig. 4.42 Sensor failures during cyclic loading of a 2nd-generation smart inlay with uncut PEI before the crack length reaches the first DAF. (Sensor 1L is excluded due to a defect after manufacture.)

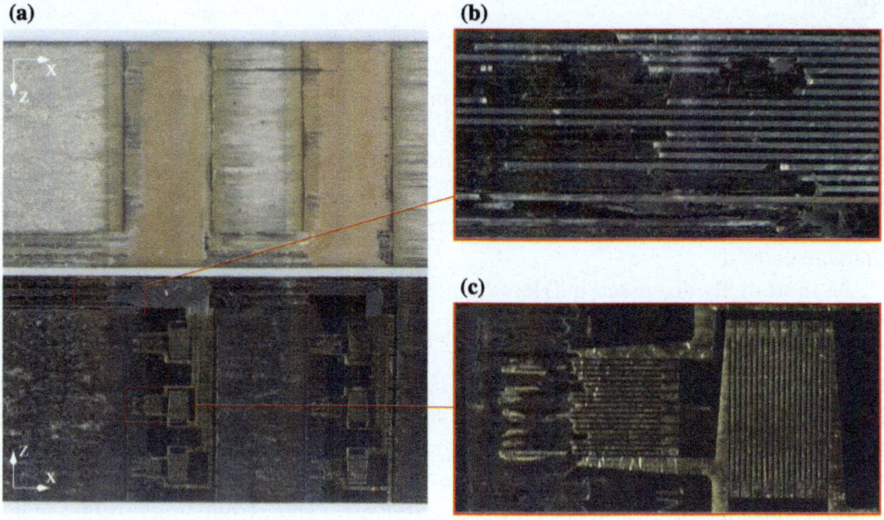

Fig. 4.43 Bondline damage in a specimen with second-generation smart inlay MDAF after a crack growth test at 3000 μm m^{-1} **a** Overview **b** Failure of the track bridge **c** Failure of the sensor grids

structures of the third-generation smart inlays, too, even though it shows a second-generation smart inlay. As soon as the crack penetrates the DAF, the sensor structures are torn apart. In conclusion, the strain capacity of the filigree sensor structures is insufficient for the strains in the bondline near the crack front that occur in the crack growth test.

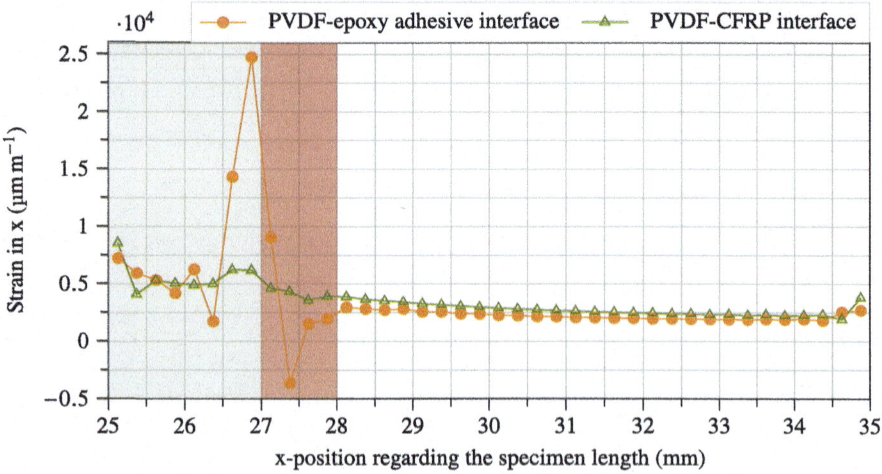

Fig. 4.44 Simulated strain in the x-direction of a PVDF DAF in the PVDF-epoxy adhesive and the PVDF-CFRP interface at a crack length of 27 mm. The grey background marks the damaged bondline and the red one the area of the strain peaks causing the sensor failures. Reproduced from [Hei+22]

The strain-to-failure upon repeated loading is the key property that determines the sensor integrity. The sensor structures of the smart inlays and the FBG fibres are very thin, which means their stiffness is low as well. Accordingly, they do not attract much of the load but have to follow the strain of their substrate. Hence, the strain values in the bondline are analysed further for a more profound understanding of the damage caused.

As thoroughly discussed in [Hei+22, pp. 7–10], an FE analysis of CLS specimens with PVDF DAFs and 3000 µm m^{-1} strap strain explains the failures of the sensors. The FE analysis reveals a major strain peak in the first millimetre of the DAF when the crack tip is right at the start of the DAF. For the PVDF-epoxy adhesive interface, the strain ε_x in x-direction of the CLS specimen peaks at 24,800 µm m^{-1}, drops to near −5000 µm m^{-1}, and rises back to approximately 3000 µm m^{-1} within that millimetre. For ε_x in the CFRP-PVDF interface, about 6200 µm m^{-1} are obtained. In that interface, the strain also does not drop to negative values, but directly approaches a value near 3000 µm m^{-1}. Figure 4.44 illustrates the different strains in the PVDF-epoxy and PVDF-CFRP interface.

The simulation is qualitatively in agreement with the FBG MDAF results. Both show a strain larger than the one introduced into the strap near the crack front. The strain values from the simulation might be exaggerating the effect, as the FBG sensor outputs at the same location are much lower, though.

The simulation also indicates strains ε_y in the y-direction of the CLS specimen at the front of the DAF if the crack has reached it. Again, the strain peaks are much higher in the PVDF-epoxy adhesive than the PVDF-CFRP interface [Hei+22, p. 8]. In contrast to ε_x, ε_y is mostly compressive.

The strain peaks at the front of the DAF agree with the damage depicted in Fig. 4.43c. It is clearly visible that the sensor damage in the front is much more severe than in the back.

The second- and third-generation smart inlays already include design improvements due to the findings from the simulation and the experience with the first-and second-generation smart inlays. The idea behind the second-generation smart inlays is a reduction of the strain peaks at the sensor structures. This is done by using the stiffer PEI substrate and moving the sensors away from the highly strained PVDF-epoxy adhesive interface towards the CFRP-smart inlay interface by placing them below the PVDF layer instead of on top of it. Moving the track bridges out of the crack growth area and the sensor structures two millimetres further away from the front of the DAFs in the third-generation smart inlays shall further reduce the strain on the sensor structures and track bridge when the crack is near the MDAF.

The fact that the first-generation smart inlays fail much sooner than the second-generation ones, indicates that the PEI substrate with PVDF cover improves the durability of the smart inlay. However, this measure is insufficient to reduce the strain on the sensor structures and track bridges below the strain capacity of these structures when they are near the crack front. Consequently, only the third-generation smart inlays are capable of crack detection in the crack growth tests because the design ensures the sensor grids and not the track bridges are the first element to be subjected to the strains near the crack front. Moreover, they are only subjected to the crack front once the crack has started penetrating the DAF. As a result, the sensors of the third-generation smart inlays can provide information on the approaching crack before they fail. However, for the sensors of the smart inlay to withstand the event of a crack travelling over them, additional refinements are necessary.

One option to improve the duration of the smart inlay sensors could be to move the sensors away from the bondline and integrate them into the laminate instead. As shown in the previous section, Fig. 4.37, strain measurements inside the adherend provide similar information, but the strain peak around the crack tip does not occur. The disadvantage is that the sensor would have to be connected through the laminate in that case, which means additional disturbances to the structure. A second option is changing the sensor material to a more robust one. In this context, screen-printed sensors using conductive pastes instead of metallic sensor grids are part of an ongoing investigation. The third option is measuring the strain in the lap only. The strain in the lap decreases instead of increasing (see Fig. 4.37). This would greatly reduce the strain on the smart inlays. The disadvantage is that the sensors might be more difficult to access. In the case of the CLS specimens, for example, the sensors could only be accessed through the laminate. Moreover, placing MDAFs in the lap contradicts the findings that no DAF is required on the load-receiving side in CLS-like configurations.

4.2.4 Influence of the Multifunctional Disbond Arrest Features on the Mechanical Strength of the Structural Bond

Is the mechanical strength of the structural bond altered by the MDAFs?

This section presents the results of the residual strength test. Moreover, it discusses additional results from the research group that provide insight into the influence of the (M)DAFs on the mechanical strength of the structural bond.

Residual Strength of the Bondline

Figure 4.45 depicts the failure load in terms of the applied force at failure for three different specimen configurations. It also shows the crack length of the specimens before the test. There are two clusters of failure loads, one around 40 kN and one around 33 kN. The difference between the load capacity of the according specimens correlates strongly with the presence of a DAF in front of the crack. All specimens, whose damage is well within the first DAF, fail around 40 kN. It does not seem to make a difference whether the crack is at the start or near the end of the DAF. In contrast, those specimens without DAF or the two-stage DAF having been passed in the crack growth test fail around the lower load of 33 kN. Hence, the DAFs in the unmodified CLS specimens have a positive effect on their strength.

The improvement in the strength of the specimens with functional DAF is evidenced by the damage type as well. All specimens with functional DAF fail in the CFRP laminate, whereas the others fail in the adhesive layer. As discussed, by Schollerer [Sch24, pp. 134–135], the surface toughening effect, which increases the bondline strength [Sch24, Sect. 4], is sufficient to shift the failure in the CLS specimens from the bondline to the laminate.

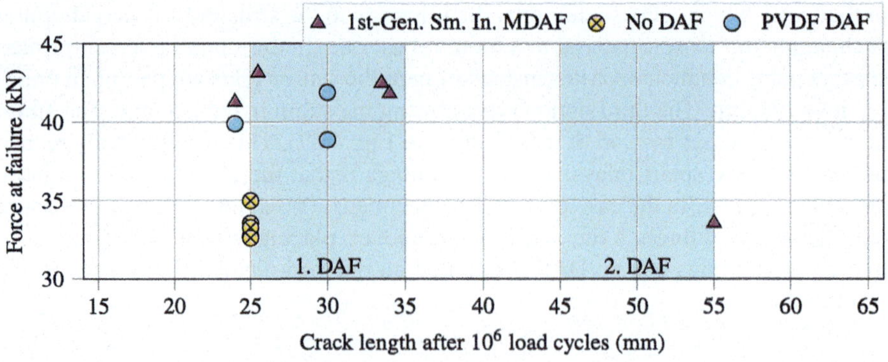

Fig. 4.45 Force at failure in the quasi-static specimens over the crack length in the specimens. The sample size of the "No DAF" configuration is 5

The surface toughening effect, in terms of redistributing the loads in the bondline, is not influenced by the applied micro strain gauges in a way the failure would occur in the bondline again. This is evidenced by the first-generation smart inlay MDAF shifting the failure into the CFRP laminate as well.

This suggests that there is an allowable disturbance by the integrated sensors in overlap joint configurations similar to the CLS joint. At or below the allowable disturbance, the global strength of the bonded structure does not decrease. Hence, the integrated sensors may weaken the surface toughening effect, but as long as it is still sufficient to shift the failure into the laminate, the disturbance is tolerable. From that perspective, the gain by minimising the disturbance due to the integrated sensors is limited by the shift of the failure mode. However, this seems to apply to static loads only because no failure in the laminate has been observed under cyclic loading.

Even though the failure in the laminate is a good indicator that the MDAFs do not jeopardise the static strength of the joint, these failures prevent a valid comparison between the different (M)DAF types. Due to the allowable disturbance, it can only be tested, if some configurations disturb the static strength of the bondline so much that the failure occurs in the bondline again. A valid comparison, however, could still help to improve the dynamic strength of the structural bond. This is the main reason why the rest of the specimens have been modified to ensure a failure of the bondline.

Figure 4.46 shows the results of the modified CLS specimens. Due to the cut in the strap, the influence of the initial crack length on the obtained force at failure is expected to be much higher. Hence, the stresses at failure are analysed as well to reduce the effect of the initial crack length. In accordance to Habenicht [Hab09, p. 412], the stresses at failure are determined by dividing the forces by the entire intact bondline at the beginning of the quasi-static test. As illustrated in Fig. 4.46, the configuration with no DAF shows the highest force but similar stresses at failure. This is because all specimens without DAF have an initial crack length of 10 mm only. This equals a remaining bondline length of 60 mm. In contrast, all other configurations include specimens with remaining bondline lengths of 45 mm or less. The first-generation smart inlays MDAFs even include a specimen that has only five millimetres bondline remaining. Hence, these show a much higher stress than force at failure and the largest standard deviation of all specimens.

From the results, there is only one clear conclusion possible. The third-generation smart inlays MDAFs show the worst force and stress at failure. This is caused by the pocket milled into the lap, which weakens the structural bond severely. For all other configurations, there is no clear trend visible in the data. One might expect a surface toughening effect that increases the strength of the structural bond, but this is not the case either.

The specimen layout, in combination with the bondline damage, explain why there is still no difference between the configurations. The second DAF is never at the overlap edge of the bond but always 15 mm away from it. The same applies to the first DAF for the non-cycled specimens. Consequently, at least one DAF is always mispositioned for the necessary load redistribution for increasing the load capacity of the bondline because it has to be placed at the overlap edge for that. Moreover,

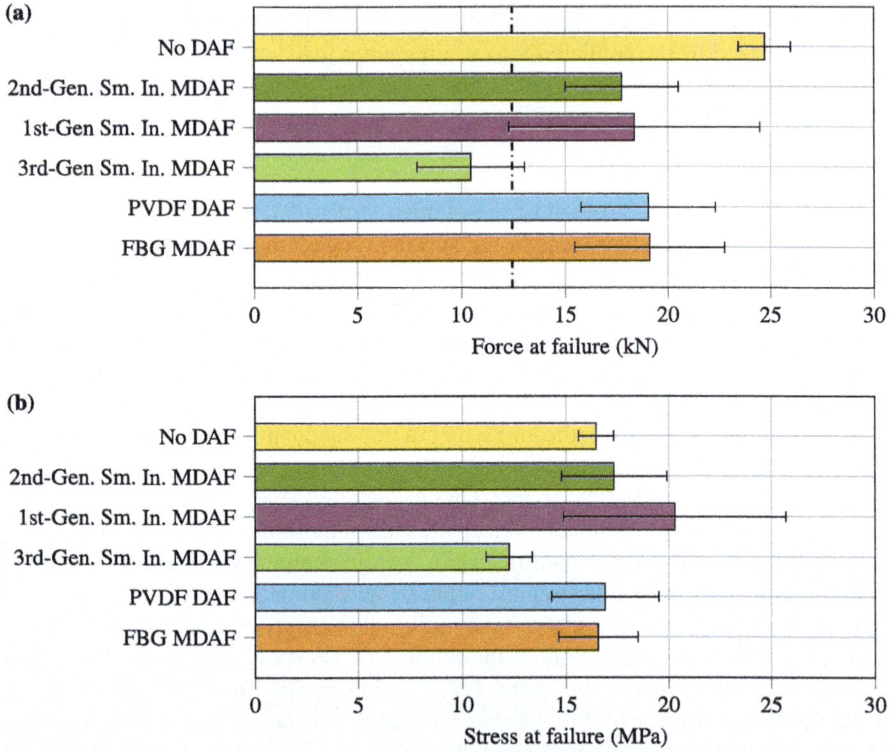

Fig. 4.46 Results of the residual strength test of the modified CLS specimens **a** Forces at failure. The dash-dotted line marks the ultimate load of CFRP **b** Stresses at failure

the specimens fail abruptly with no visible crack growth that could be arrested by the DAFs. Instead, the DAFs are pulled out of the CFRP laminate nearly the same instant the bondline fails. As a result, the specimens with DAF behave similarly to those without in that test. In conclusion, the residual strength tests with modified CLS specimens are not suited for evaluating the effect of the (M)DAF configurations on the static strength of the bondline. Obtaining the energy release rates for mode I and II using Double Cantilever Beam (DCB) and End Notched Flexure (ENF) tests, for example, seems more appropriate for a more profound understanding and quantification of the influence of the (M)DAFs on the mechanical properties of a bondline.

There are already results in the literature that allow for some evaluation of the influence of the (M)DAFs. These results come from partner projects of Schollerer [Sch24] and Kyriazis et al. [Kyr+21], who investigate the surface toughening effect and PEI-foil sensors for the integration into CFRP laminates, respectively. The results are of high relevance for the (M)DAFs because the materials for the PVDF DAF and PEI sensors are identical to those of the (M)DAFs. Schollerer's investigations can be used directly to evaluate the influence of the PVDF DAFs on the mechanical properties because the same materials are used for the CFRP laminate and DAF. The differences between the PEI-sensors of Kyriazis et al. and the second- and third-generation smart inlays are the geometry and additional PVDF cover. Moreover, Kyriazis et al. investigate foil sensors integrated into a CFRP laminate. Hence, they do not consider an additional adhesive layer. This is why, these results cannot be applied directly to analyse the influence of the MDAFs on the mechanical properties, but for some effects.

Schollerer [Sch24, pp. 92–101] finds that DCB specimens with PVDF DAF increase the critical energy release rate in mode I by 25.1 % in the area of the PVDF [Sch24, pp. 92–97]. Shortly before the crack reaches the DAF the critical energy release rate decreases slightly. Moreover, Schollerer reports a failure in the CFRP-PVDF interface, which is similar to the critical damage path in the CLS specimens (see Fig. 4.17). An improved CFRP-PVDF interface could thus further improve the mode I behaviour.

Schollerer [Sch24, pp. 97–101] studies ENF tests with PVDF DAF as well. In this case, the DAF leads to a complete disbond arrest, which prevents the determination of the critical energy release rate in mode II [Sch24, pp. 97–102]. Again, the crack growth rate elevates before the DAF in comparison to the reference.

Schollerer's results illustrate that the critical energy release rates correlate with the disbond arrest effect of the surface toughening material. As expected, in the area of the DAFs, there is a local increase in the critical energy release rate, which explains why the cracks arrest. The decrease in the critical energy release rate when the crack is near the DAF shows the possibility of a negative influence of the DAF on the static strength of the structural bond. For crack distances at which the critical energy release rate is reduced, the static load capacity would decrease as well. However, the static load capacity in the residual strength test with modified CLS specimen is still well above the ultimate strain of CFRP structures ($4000\,\mu m\,m^{-1}$) for most specimens, even though the strap is cut and the DAF is mispositioned (see Fig. 4.46). The exceptions are the specimens, whose lap is damaged severely or whose second DAF has been breached. Hence, the scenario that a pre-damaged bondline fails statically because of the PVDF DAF seems unrealistic. For dynamic crack growth, the elevated crack growth rate is even less critical, as the crack is arrested afterwards.

The second important question is how the embedded micro sensors affect the mechanical properties. Kyriazis et al. [Kyr+21] study interlaminar shear specimens regarding the influence of a PEI foil and additional micro sensor structures made of gold on the mechanical properties in mode I and II. They find that the PEI foil increases the critical energy release rate in mode I by about 100 %. Adding gold structures to the foil reduces the critical energy release again, but in combination with the PEI, there is an improvement compared to the reference without integrated PEI foil and sensor [Kyr+21, p. 16]. Moreover, Kyriazis et al. [Kyr+21, p. 14] demonstrate that the Interlaminar Shear Strength (ILSS) of a CFRP laminate does not change due to the presence of the PEI, whereas the gold structures decrease it. The area share of the gold structures relative to the area of the PEI foil plays an important role in that context [Kyr+21, p. 14]. Area shares above 50 % change the damage mode from interlaminar failure in the CFRP to a failure in the PEI-sensor-epoxy interface because of the weak gold-epoxy interface. Kyriazis et al. [Kyr+21, p. 18] conclude that an area share below 30 % is sufficient for ensuring the gold sensor structure does not reduce the ILSS. They explain this by the support effect of the strong PEI-epoxy interface around the sensor structures.

The negative influence of the gold-epoxy interface is expected to apply to the smart inlays as well. The failure of the similar gold-epoxy adhesive interface of first-generation smart inlay MDAFs, which is depicted in a previous section, Fig. 4.18a, strengthens that assumption. Since a failure in the PVDF-epoxy adhesive interface is not observed in the other two generations, the gold-PVDF interface seems to be a potential weakness in the structural bond. Moreover, according to the CLS specimens with second-generation smart inlays, the combination of PEI-PVDF and gold-PVDF interfaces is incapable of providing the required adhesive strength for the disbond arrest function.

The support effect, thus, seems to be key for the disbond arrest function of the second- and third-generation smart inlay MDAFs. This is evidenced by the findings in Sect. 4.2.1 that cutting the PEI substrate around the sensor structures provides a disbond arrest function of these MDAFs. However, there is no data to quantify the strengths of the PEI-PVDF and gold-PVDF interfaces.

The possibility to vary the sensor area and to support weak interfaces by neighbouring strong interfaces indicates that there are many design options to ensure a functionally compliant multifunctional bondline. A profound understanding of the complex interaction between the combined interfaces is, thus, important. For that matter, it seems appropriate to characterise the individual and combined interfaces in future works. This data could also serve as input parameters in an FE model of the multifunctional bondline for crack growth predictions later on.

Effect of the Integrated Sensors on the Laminate and Bondline Geometries

The fibre orientation and the bondline thickness are two crucial parameters for the strength of bonded composites. A misaligned fibre orientation reduces the compression strength of a composite, which is why fibre waviness has to be avoided to obtain a high compression strength [Sch07, p. 355]. Habenicht [Hab09, pp. 442–446] illustrates how the bondline thickness has a great influence on the bondline strength. Bondline thicknesses between 50 μm and 200 μm provide the maximum strength. Hence, the influence of integrated MDAFs on these parameters is discussed in the following by microsections of an FBG MDAF and a first-generation smart inlay MDAF.

Figure 4.47 depicts the microsections of an integrated FBG fibre within and outside the MDAF. From Fig. 4.47a it is visible that the co-cured FBG MDAF causes a severe deflection of the 0° ply below the FBG fibre. Moreover, a slight deflection of the plies below is visible, too. Outside the MDAF, the co-cured FBG fibre penetrates the 0° ply and seems to push the carbon fibres aside (see Fig. 4.47b). This also causes deflections of the surrounding plies.

Both images in Fig. 4.47 further illustrate that the bondline thickness is influenced in the area surrounding the FBG fibre. In the MDAF, resin pockets are visible, which increase the bondline thickness noticeably. This effect is much less outside

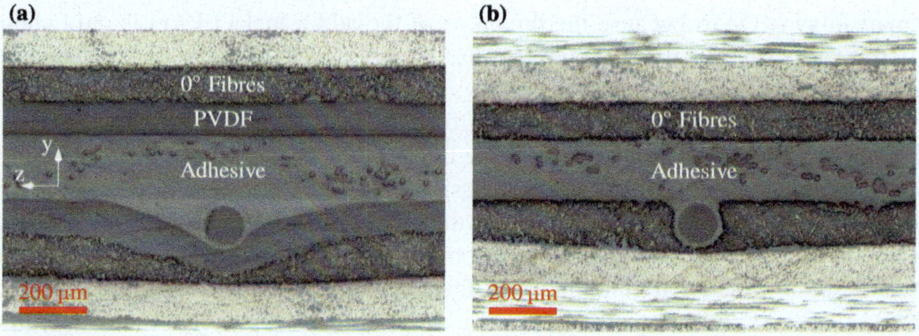

Fig. 4.47 Microsection of a co-cured FBG fibre **a** Cross-section in the FBG MDAF **b** Cross-section outside the FBG MDAF

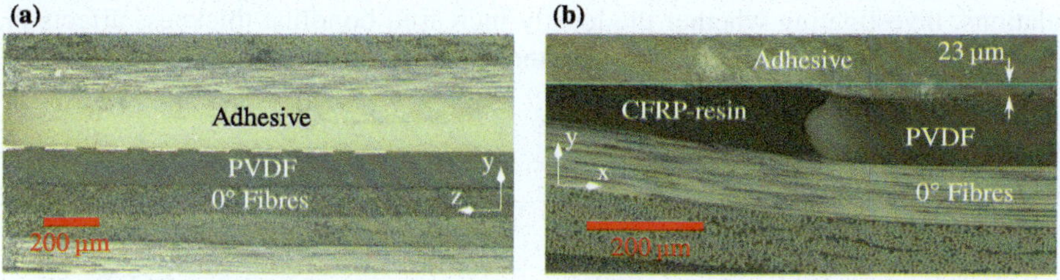

Fig. 4.48 Microsection of a first-generation smart inlay MDAF **a** Cross-section showing the track bridge in front of the MDAF **b** Cross-section of the MDAF showing an increased bondline thickness above the PVDF. Previously published in [Hei+21, p. 11]

the MDAF. This illustrates that the PVDF preventing the FBG fibre from penetrating the 0° ply of the CFRP causes a larger deviation of the bondline thickness instead. In both images, the bondline thickness away from the FBG fibre is approximately the same. Hence, the influence seems to be limited to the area around the FBG fibre only.

Figure 4.48 shows two microsections of first-generation smart inlay MDAFs. Figure 4.48a depicts the track bridges of the MDAF outside the DAF area. In contrast to the FBG fibre, there is no significant deflection of the fibres visible. Figure 4.48b depicts a different plane, which shows the start of the DAF area. In this cut, there are no sensor grids visible. Instead, it illustrates that the co-cured PVDF not only deflects the fibres and causes a CFRP-resin pocket, but also alters the bondline thickness locally. The bondline thickness increases by approximately 23 µm above the PVDF. This might be caused by a different thermal expansion of the CFRP resin and the PVDF.

The microsections show that both the fibre orientation and the bondline thickness are altered by the co-cured MDAFs. The 150 µm-thick FBG fibres show a more severe distortion than the sensors of the first-generation smart inlay MDAF, whose thickness is about 7 µm in the area of the track bridge and 200 nm at the sensor layers [Hei+21, p. 6]. Similar behaviour is expected for the second- and third-generation smart inlay MDAFs because the thickness of the additional PEI foil is only about 10 μm. Accordingly, the FBG fibres cause larger deviations in the fibre orientation and bondline thickness around them.

From the residual strength and crack growth results, no direct correlation between the altered fibre orientation and bondline thickness is possible. The fact that disoriented and wavy fibres affect the compression strength, in particular, means that the tensile tests do not suffice to prove that the co-cured MDAFs do not influence the static strength of the laminate. This emphasises the need for further investigations to assess the functional compliance of the structural bonding, disbond arrest, and SHM functions regarding other load cases. About the local increase in the bondline thickness above the PVDF, it appears that there is a similar gap in the microsection of Schollerer [Sch24, p. 85] in a specimen with PVDF surface toughening. Since the surface toughening effect is still not precisely reproducible in numerical simulations, investigating whether the locally increased bondline thickness affects the surface toughening effect might be of interest.

4.2.5 Conclusions

All PVDF-based DAFs, whose PVDF is in direct contact with the CFRP, show a disbond arrest function. In contrast, the PEI DAF and second-generation smart inlay MDAF with uncut PEI are incapable of arresting the crack throughout the entire crack growth test. While the PEI DAF does not even slow down the crack growth, the second-generation smart inlay MDAFs with uncut PEI temporarily stop the crack. However, the untreated PEI-PVDF interface is incapable of withstanding the loads at the crack tip for long and thus fails eventually during cyclic loading. The results show a few failures of PVDF-based DAFs during the crack growth test as well. These have a failure of the untreated co-cured CFRP-PVDF interface in common. The crack growing through this interface is thus considered the critical crack path. It is not exactly certain whether the failure of the interface is adhesive or special cohesive, but the fact that no failure of the plasma-pre-treated PVDF-epoxy adhesive interface is observed suggests that the reliability of the PVDF-based DAFs could be improved by a surface activation of the PVDF before co-curing.

The strain gradient is a possible damage indicator for overlap joints. In the case of the CLS specimen, even two measurement points at one side of the overlap suffice for identifying a crack that has grown to the two-stage MDAF if they are slightly more than the load transfer length apart. Subtracting the sensor signals and introducing a threshold appears to be a simple way for damage identification. However, the difference is not only sensitive to damage but also the applied load, which complicates finding a proper threshold. Another option is analysing the signal quotient. For high loads, this criterion is much less sensitive to load changes. Even though a precise determination of the crack length is not possible with the tested arrangement, it seems feasible to estimate the remaining distance from the crack to the MDAF to some extent by using the signal quotient. However, the quotient criterion appears unreliable if the sensor outputs are low. Moreover, the well-known conditions of the CLS specimen mean that further investigations with more complex damage phenomena are necessary to ensure a reliable damage detection in practical applications.

The FBG sensors show the capability to withstand the high strains near the crack tip in the structural bond. In contrast, the foil sensors are incapable of withstanding these strains repeatedly. However, they have shown sufficient capacity to withstand the strains before the crack has reached them, which enables detecting the event of the crack growing near the MDAF.

The residual strength tests with unmodified CLS specimens demonstrate that PVDF-based DAFs improve the static strength sufficiently to shift the failure into the laminate. The modified CLS specimens do not provide an insight into the disturbance of the mechanical strength due to the integrated (M)DAFs. Results from similar experiments within the research group, however, show that the PVDF DAFs reduce the critical energy release rate when the crack is near, but significantly improve it at the DAF. Moreover, there is clear evidence that the PEI substrates of the smart inlay do not impair the mechanical strength but even double the critical energy release rate under mode I loading. In contrast, the gold sensor structures of the smart inlays show a weak interface, which weakens the strength of the bondline in modes I and II. However, the PEI is capable of compensating for this effect if the area share of the sensor structures is below 30 %. Moreover, the microsections of integrated MDAFs illustrate that both the fibre orientation and the bondline thickness are affected differently by the co-cured first-generation smart inlay and FBG MDAFs.

Regarding Research Hypothesis 1 the results support the following conclusions: The integration of MDAFs into epoxy bondlines enables the disbond arrest and SHM functionalities without reducing the mechanical strength of the bond significantly as long as the weak interfaces, such as the gold-epoxy or gold-PVDF interfaces, are supported substantially by surrounding strong interfaces, such as PVDF-epoxy or PEI-epoxy. After disbond arrest, the MDAFs even improve the static strength of the structural bond. Regarding the system boundary of the CLS specimen, the aspect that embedding MDAFs into epoxy bondlines at discrete positions enables the disbond arrest and SHM functions while maintaining the mechanical strength of the bondline is confirmed. For the aspect of maintaining the weight-saving potential of an ordinary healthy bondline, it can only be said that the MDAFs themselves consist of lightweight materials in contrast to a mechanically fastened joint. However, there is no substantial conclusion regarding the weight-saving potential possible from the discussed tests on the laboratory scale because the major weight contribution is expected due to the required measurement equipment and sensor connections (see Chap. 5).

In terms of Research Hypothesis 1.1, it can be added that the integrated sensors show unambiguous signals in correlation with the crack length. However, only the FBG sensors do not fail upon repeated mechanical loading. The foil sensors of the smart inlays do not show sufficient strain capacity to withstand repeated mechanical loading when they are near the crack tip in the bondline. Changing the sensor design to keep them away from the high strains around the crack tip and the PVDF-epoxy adhesive interface enables them to withstand the repeated straining before the crack has reached the smart inlays, though. Hence, Research Hypothesis 1.1 can be confirmed for the FBG sensors, but not the smart inlays for the given load conditions.

The results confirm Research Hypothesis 1.2 as well. For the system boundaries of the CLS specimen, the strain gradient in the MDAFs enables a clear judgment once the damage reaches it. It seems feasible to apply a similar system in practical applications as well. However, a profound understanding of the load conditions in practical applications is required to ensure clear damage detection. Depending on the system, this might require more sophisticated evaluation algorithms than the two discussed ones. Moreover, it should be evaluated if additional sensors, e.g. at both sides of the overlap joint, are required for a clear judgement.

4.3 Standard Curing Processes

Is manufacturing the multifunctional bondline with standard curing processes feasible?

This section presents the results from the experimental investigation regarding the effect of the transient material properties of PVDF during curing. It discusses the integrity of the sensor structures during curing first, before dealing with the feasibility of maintaining the piezoelectric effect of PVDF despite melting it due to the hydrostatic pressure in an autoclave.

4.3.1 Impact of the Curing Process on the Sensor Structures of PVDF Film Sensors

What is the impact of the curing process on the PVDF film sensors, and how to avoid damage to the sensor structures during curing?

This section deals with the sensor integrity of the PVDF foil sensors for the first-generation smart inlay MDAFs during curing. The focus is on the integrity of the sensor structures upon melting of the PVDF material, and the influence of the surrounding materials on the sensor integrity.

Figure 4.49 shows two image series taken with a USB microscope while annealing PVDF film sensors on a hot stage. The images show the sensor structure during heating from 150 °C to 180 °C, and cooling down to 150 °C again. As the melting temperature of the α- and β-phase of PVDF lies between 167 °C and 172 °C [MLL14, p. 687], the material melts in the process. The film sensor in Fig. 4.49a has been placed on a microscope slide and fixed to it at one short edge only to allow for a relatively

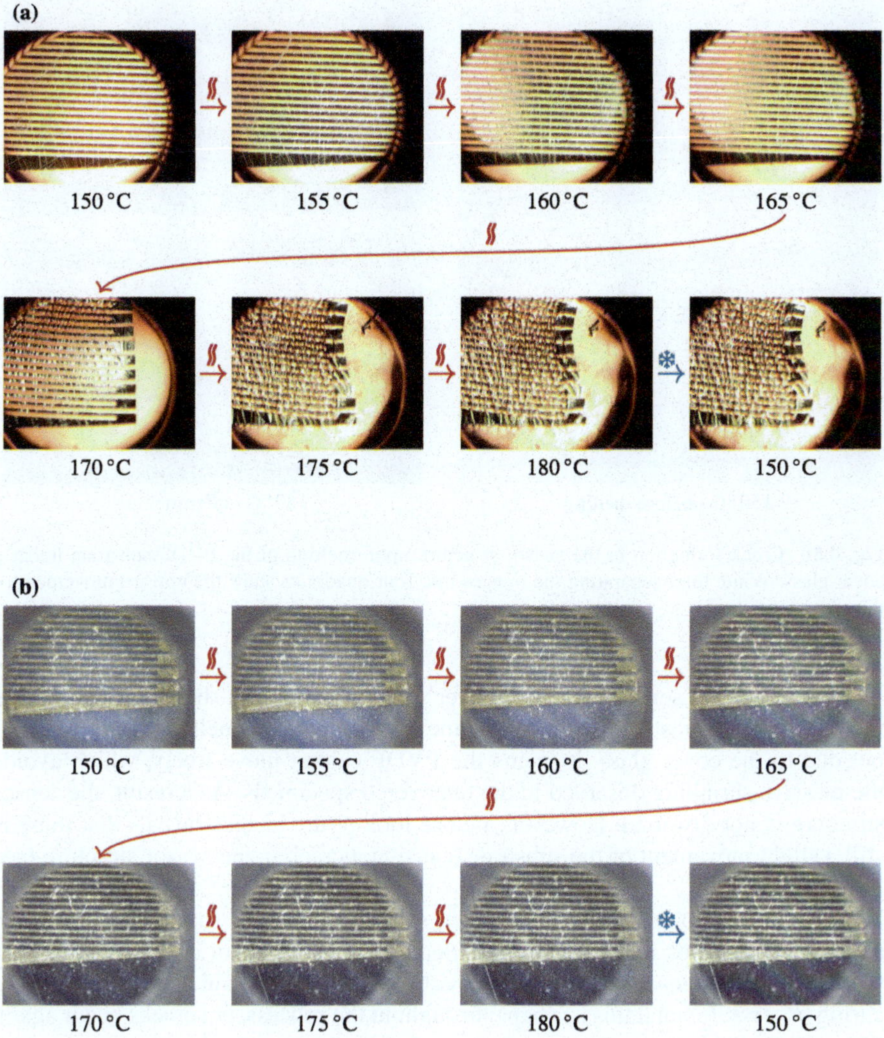

Fig. 4.49 Annealing a PVDF film sensor under a microscope **a** Film sensor on a microscope slide without fixation. Sensor structures distort **b** Film sensor on a microscope slide fixed by a cover glass. No visible distortion

free movement of the substrate. In contrast, the entire sensor structure of the film sensor in Fig. 4.49b is fixed by a cover glass.

The uncovered sensor structure distorts massively when the substrate melts, and does not recover upon cooling down. The cause is the PVDF substrate that starts moving and shrinking when reaching the melting temperature. The sensor structure cannot withstand that movement, which results in the distortion depicted in Fig. 4.49a.

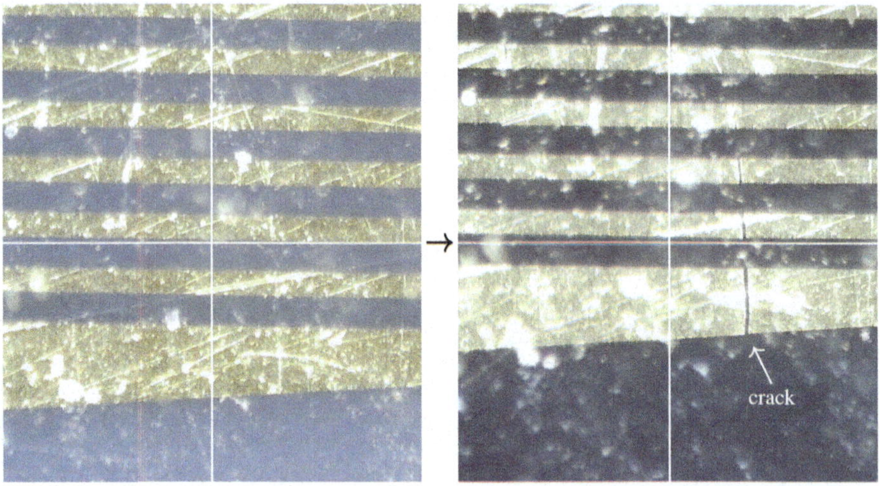

Fig. 4.50 Crack formation in the sensor structure upon melting of the PVDF substrate under a cover glass. White lines separating the images into four quadrants show the upward movement of the sensor structure

In contrast, the sensor under the cover glass maintains its shape throughout the melting process, mostly. There is still some movement in the melted PVDF material, but due to the cover glass, it seems the PVDF cannot move freely, which avoids the massive shrinking observed in the uncovered specimens. As a result, the sensor structure is not distorted. However, a close look at the images reveals that there is still a slight movement of the substrate, causing a crack in the sensor structure (see in Fig. 4.50[26]).

The effect of co-curing the PVDF film sensors to a CFRP structure is similar to annealing in the hot stage. Figure 4.51 depicts a PVDF sensor after co-curing to a CFRP plate. The sensor structure is in direct contact with the caul plate during the co-curing process. It maintains its shape throughout the process, but cracks occur again. Figure 4.52a shows a single row of a smart inlay with three half bridges covered by another PVDF strip. In contrast to the PVDF sensor without a PVDF cover, severe damage to the sensor structure occurs throughout the co-curing process. A comparison to a similar sensor without cover layer, as shown in Fig. 4.52b, demonstrates the severity of the damage. The film sensor in Fig. 4.52b has been placed upside-down

[26] The images are the first and last image of the image series in Fig. 4.49.

Fig. 4.51 Crack formation in a PVDF film sensor upon co-curing the sensor to a CFRP plate with the sensor structure being in contact with the caul plate

Fig. 4.52 PVDF film sensors, whose sensor structures are not in direct contact with the caul plate upon co-curing **a** Sensor structure of a PVDF film sensor covered by another layer of PVDF **b** PVDF film sensor flipped over so that the sensor structure is in contact with the CFRP prepreg

to test if contact with the laminate instead of the caul plate is sufficient to prevent the massive distortion of the sensor structures as well. As illustrated, this turns out to be true. In conclusion, the PVDF film sensors without PVDF cover behave similarly to those on the hot stage with fixation by the cover glass. The film sensor with PVDF cover behaves similarly to the unfixed one on the hot stage instead.

The results demonstrate the importance of providing the support function for the foil sensor structures by the surrounding material. Without the support function, the movement of the PVDF upon melting damages the sensor structures massively. However, the crack formation upon melting of the PVDF substrates illustrates that providing the support function is a necessary, but not a sufficient condition, for the successful integration of PVDF-based film sensors.

Von der Heide et al. [Hei+21] find that the fabrication of the film sensors itself has a big impact on the integrity of the sensor structure upon curing, too. The main challenge in preparing the film sensor is providing a smooth foil surface before adding the sensor structures to ensure electrical connections [Hei+21, p. 13]. Von der Heide et al. [Hei+21, p. 9] illustrate how insufficient preparation of the PVDF substrate even causes defects in the sensor grid before curing. Considering that the film sensors

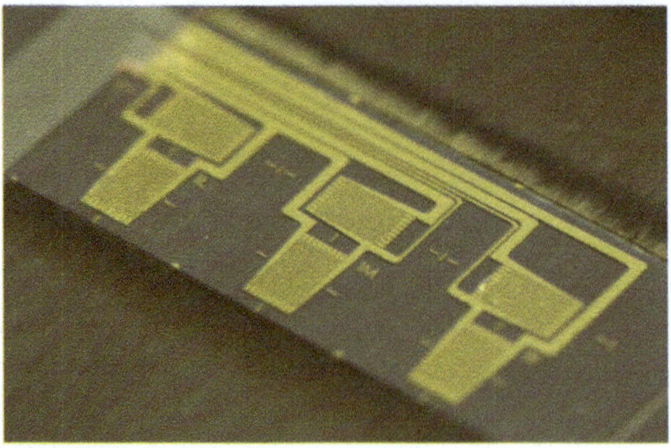

Fig. 4.53 Successfully co-cured first-generation smart inlay with smoothened PVDF substrate and direct contact of the sensor structures to the caul plate. Previously published in [Hei+21, Fig. 11a]

depicted in Figs. 4.49, 4.50, and 4.51 have been made without the necessary smoothing, it is likely that an insufficient pretreatment of the PVDF substrate contributes to the crack formation [Hei+21, p. 12].

Providing the support function in combination with the pretreatment of the PVDF substrate enables co-curing PVDF film sensors to CFRP to some extent. Figure 4.53 shows the first DAF sensors of a first-generation smart inlay with six functioning sensor grids after co-curing. Out of 360 sensor grids of this type, approximately 90 % have been functioning after smart inlay manufacturing, and 70 % after co-curing [Hei+21, p. 12]. This demonstrates that co-curing PVDF film sensors to CFRP parts using standard curing processes is feasible if the surrounding material provides the support function, but is still very demanding.

4.3.2 Phase Transformations of Polarised PVDF upon Annealing in an Autoclave at Various Maximum Temperatures

How do the piezoelectric phase contents of polarised PVDF change upon annealing in an autoclave at 3 bar pressure and various maximum temperatures?

This section summarises the results obtained by Vasic et al. [Vas+21] regarding the effect of annealing polarised PVDF in an autoclave. In the scope of this thesis, the most relevant results for testing Research Hypothesis 2.2 are illustrated. These are the crystallinity and the share of piezoelectric phases of the PVDF. For further details, please refer to [Vas+21].

Figure 4.54 shows the total crystallinity and the fraction of electroactive phase before and after annealing. The results indicate that the total crystallinity tends to drop slightly from about 55 % to around 50 %. At 160 °C the crystallinity tends to

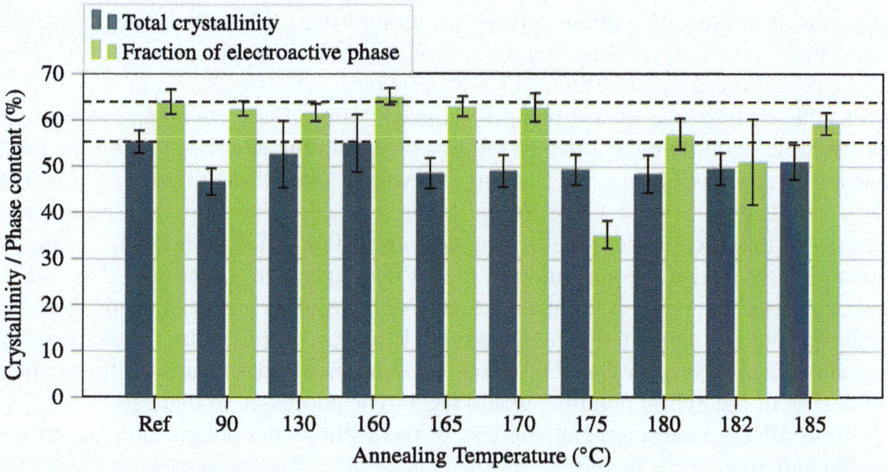

Fig. 4.54 Comparison of the total crystallinities X_c and fraction of electroactive phases F_{EA} before (Ref) and after annealing at different temperatures

be maintained, but there is a rather high standard deviation of ≈10 %. The biggest drop of crystallinity occurs at an annealing temperature of 90 °C. As for the fraction of the electroactive phase, the samples annealed below the melting point show a different behaviour than those annealed above it. The melting point of the reference is 171.2 °C [Vas+21, p. 5]. For the temperatures below, the fraction decreases slightly or remains the same. Again, for 160 °C the highest fraction of electroactive phase is observed by a small margin. Slightly above the melting point at 175 °C, there is a significant drop. For even higher temperatures, the fraction of the electroactive phases becomes significantly less. However, all samples with annealing temperatures above the melting point show a bigger loss in the fraction of the electroactive phase than those samples annealed below the melting point.

As discussed in detail in [Vas+21, p. 5], the DSC analyses of all samples annealed at temperatures above the melting point, i.e. 175 °C or higher, show the formation of γ-phase. The samples annealed at 175 °C and 180 °C also show γ'-phase contents. In contrast, the other samples only show α- and β-phase contents.

The results demonstrate that the autoclave process at 3 bar pressure has not increased the β-phase content. Thus, the observation of Nakamura et al. [Nak+03, p. 1704] that annealing at up to 182 °C while keeping the length of the PVDF constant leads to β-crystals only is not reproducible with the presented process. Instead, the results suggest that the γ-phase is formed out of the β-phase. However, it cannot be said with certainty that no β-phase is left after exceeding the melting temperature.

Since the piezoelectric coefficient of γ-phase PVDF is about five times less than the one of β-phase PVDF [MLL14, p. 696 and p. 688], it is questionable that the piezoelectric effect of γ-phase suffices for monitoring the bondline integrity. However, the results also demonstrate the feasibility of utilising the piezoelectric effect if the process temperatures are at 170 °C or below.

In this context, the hybrid bondline concept comes into view again, as it might be possible to co-cure the outer PVDF layers by melting and then joining the inner sensor layers (and electrodes) to them afterwards without melting (see Fig. 3.7b). This could be achieved by a diffusion bonding process, for example. The decisive factors in this process are whether the adhesion after diffusion bonding meets the requirements, and if a polarisation of the PVDF is possible afterwards. This seems particularly challenging if the shear effect shall be used because this would require a polarisation orthogonal to the field direction [Sin21, p. 43]. Besides the challenge of providing a piezoelectric PVDF material, the tolerance issues regarding the bondline thickness of the hybrid bondline would have to be addressed in that case.

If the PVDF material melts, the loss of the β-phase and polarisation are not the only challenges for utilising the piezoelectric effect. The movement of the PVDF upon melting might cause the electrodes to break, to lose their alignment or to get in touch with each other. Moreover, the thickness of the PVDF material could change. All of these would either break the sensor or change its performance significantly. Hence, a way would be required to ensure the geometry of the PVDF sensor stays within the design tolerances in addition to preserving its phase.

4.3.3 Conclusions

The investigations of the PVDF foil sensors in the hot stage and autoclave illustrate that the PVDF not only loses its support function for the sensor structure but even causes high deformations of the sensor structure upon melting. Without another material supporting the sensor structures, they distort severely and, thus, lose their functionality. A rigid material like a cover glass or caul plate can provide the support function and protect the sensor structures. It appears even the epoxy resin has cured sufficiently when the PVDF melts to take over the support function. At the same time, the results demonstrate that encapsulating the sensor structures completely in PVDF is not possible without breaking the sensors if the PVDF melts during curing.

The support function is a necessary but not a sufficient condition for ensuring the sensor integrity during curing. The PVDF foil requires smoothing before the sensor structures are applied to avoid crack formations during curing as well.

The annealing experiment of polarised PVDF in an autoclave demonstrates that the autoclave pressure of 3 bar is insufficient to increase the β-phase content. Instead, it seems more likely that the β-phase converts to γ-phase upon melting of the PVDF. For temperatures at 170 °C and below, there seems to be sufficient crystalline β-phase for a piezoelectric sensor, though. Hence, it might be possible to use a diffusion bonding process for building a piezoelectric hybrid bondline.

The findings lead to the following conclusions regarding Research Hypothesis 2 and its sub-theses: Research Hypothesis 2.1 is confirmed, as the caul plate and cover glass are capable of protecting the foil sensor structures against the load applied to them upon melting of the PVDF. In contrast, Research Hypothesis 2.2 cannot be confirmed because the β-phase is very likely converted to γ-phase upon melting of the PVDF. Accordingly, the results also confirm only parts of Research Hypothesis 2.

4.4 System Complexity

How do the competitive and complementing functions affect the system complexity of the multifunctional bondline?

This section takes on the results from the previous sections for discussing how the complementing and competitive functions affect the system complexity.

4.4.1 Complexity Reduction by Lowering Individual Requirements of Complementing Functions

Is it feasible to decrease the system complexity by lowering the individual requirements of the disbond arrest and SHM features without compromising the safety of the structural bond?

This section analyses complementing aspects between the disbond arrest and SHM functions. The focus is on evaluating the feasibility of simplifying the individual features in comparison to a system consisting of either a disbond arrest or an SHM feature.

The quasi-static tests illustrate the complementing characteristics of the two features regarding the requirement not to decrease the static strength of the bondline. The findings from Sect. 4.2.4 that the (M)DAFs shift the failure from the bondline to the laminate illustrate that the disturbance of the DAF by the integrated sensors is negligible as long as the failure still shifts to the laminate. Only if the disturbance impairs the surface toughening effect of the DAF so much that the failure occurs in the bondline again, the integrated sensors would affect the static strength. This means there is a margin of acceptable disturbance due to the sensor integration. Hence, for configurations similar to the CLS one, the minimum requirement defining the allowed disturbance of the integrated sensors could become less strict regarding static loads.

The crack growth results show the complementing nature of the disbond arrest and SHM features directly. While the sensors seem to have a slightly negative effect on the disbond arrest function, they do not jeopardise it. Instead, they provide valuable information on the bondline status. It is especially the SHM function, which profits from the combination with the DAF, then.

This is evidenced by the proposed difference measurement in combination with a threshold value for damage identification. Looking for a universal threshold for the CLS specimens shows that the exemplary threshold of 200 μm m^{-1}, identified for the specimen in Fig. 4.34, is not suitable. It would work for specimens tested at 3000 μm m^{-1} strap strain maximum as well. However, at the lowest load of 0.928 kN in this setup, the threshold is exceeded at 24 mm only. In contrast, for the specimen tested at 4000 μm m^{-1} with third-generation smart inlay MDAFs, the threshold would be too low and trigger at the beginning of the test at 10 mm crack length for the maximum load of 12.43 kN. Doubling the threshold would work for this and the other specimens as well. However, for the FBG MDAFs, the crack would have to reach the DAF for the threshold to be exceeded. Due to the disbond arrest function, this appears to be sufficient because the crack will stay there long enough to ensure it is indicated in time. Consequently, the presence of the disbond arrest function would allow for a less conservative threshold that activates at larger damages.

In addition, the DAF allows for repeated measurements at the same condition, which could be used to increase the robustness of the damage detection. This is even evidenced by the specimens with second-generation smart inlay MDAF and uncut PEI, which does not arrest the crack throughout the entire test. As discussed in Sect. 4.2.1, it slows down the crack growth so that there is 175,000 more cycles' time before the two-stage MDAF is passed. Between the two events of the crack reaching the MDAF (a length of 25 mm) and passing it completely (a length of 65 mm), there are about 300,000 cycles in comparison to 100,000 cycles in the specimen without DAF. Hence, there is much more time for identifying the damage, even though the DAF does not even function properly. This suggests that it may be feasible to trigger an alarm, not the first time the threshold is exceeded, but only after it is exceeded multiple times. This way, false alarms could be avoided without putting the structure at risk.

As a result, it can be confirmed that it is feasible to lower the requirement of the SHM system from a precise damage characterisation to the identification of damaged bondline sections.

Regarding the complexity reduction of the DAF function, the picture is not as clear. On the one hand, the crack growth tests (see Sect. 4.2.1) indicate that the SHM system can detect damage before it reaches a critical size, even if the DAF fails eventually. On the other hand, the results suggest that the surface toughening effect of PVDF is capable of withstanding cyclic loads sufficiently to provide the disbond arrest function throughout the entire crack growth test. Considering the very high load conditions, with limit load and above, the surface toughening technology shows potential for a reliable disbond arrest technology without the additional SHM system already. From that perspective, the complexity reduction regarding the manufacturing process by utilising surface toughening instead of the hybrid bondline seems feasible without the SHM system.

However, the crack growth results only include high cyclic loads, but neither any other kind of bondline degradation nor changing environmental conditions. To be certain that the surface toughening effect works in the long term in the harsh environment of aircraft structures, additional investigations are required.

Schollerer [Sch24, Chap. 8] reports that exposing a bondline with PVDF surface toughening elements to 75 °C and 85 % humidity for multiple days or changing the ambient temperature, negatively affects the static strength of the structural bond. Regarding the exposure to heat and humidity, the strength drops by about 20 % after 1000 h compared to the unexposed specimens with surface toughening. In comparison to unexposed specimens without surface toughening, the strength is still about 60 % higher, though [Sch24, p. 146]. This changes dramatically for operating the bondline below the glass transition temperature of PVDF around −40 °C to − 30 °C [Ame09, p. 6635]. In an ambient temperature of −55 °C the strength with surface toughening drops to 58 % in comparison to the specimens without surface toughening at room temperature [Sch24, p. 147]. For the specimens without surface toughening, the strength drops to 65 % only, so the PVDF surface toughening even decreases the bondline strength in this condition. As a result, PVDF is not suited as a surface toughening material for ambient temperatures below its glass transition temperature. Even though Schollerer investigates surface toughening for increasing the static strength and not for disbond arrest, the loss of the surface toughening effect is likely true for a DAF as well. The SHM system cannot change that, which means for applications at this temperature, another material is required that can provide a surface toughening effect. In this context, it might be beneficial to decrease the requirements of that material to slow down damages instead of fully arresting them due to the presence of the SHM system.

Reducing the requirement from avoiding fatigue failures to providing stable crack growth enables the usage of other DAFs as well. For example, the concave/convex DAF from Hirakawa et al. [HYM22] seems considerably easy to manufacture and increases the fracture toughness, so that cracks are slowed down significantly. However, it does not provide a disbond arrest function. The combination with the proposed SHM method could help qualify such DAFs, nevertheless. Enabling more solutions regarding the DAF function increases the solution space for the entire system. As a result, solutions with a satisfactory DAF function but significant benefits in terms of weight or manufacturability, for example, might be found.

In this context, the damage tolerance approach discussed in Sect. 3.2.3 comes into view again. According to this, the requirements for the MDAF demand that the inspection intervals for either slow growth or arrested growth provided by the SHM system are short enough to detect damage before it reaches the critical size. The DAF thus needs to guarantee damage grown to the detectable size does not grow to the critical size in less than the provided inspection interval. This seems possible by disbond arrest or slowing down further damage growth via the MDAF at damage sizes big enough for clear detection, but small enough for sufficient residual static strength of the remaining undamaged bondline. As illustrated schematically in Fig. 4.55, the two-stage MDAF is capable of fulfilling this demand. According to the crack growth test results, the DAFs provide the demanded disbond arrest, while the SHM feature

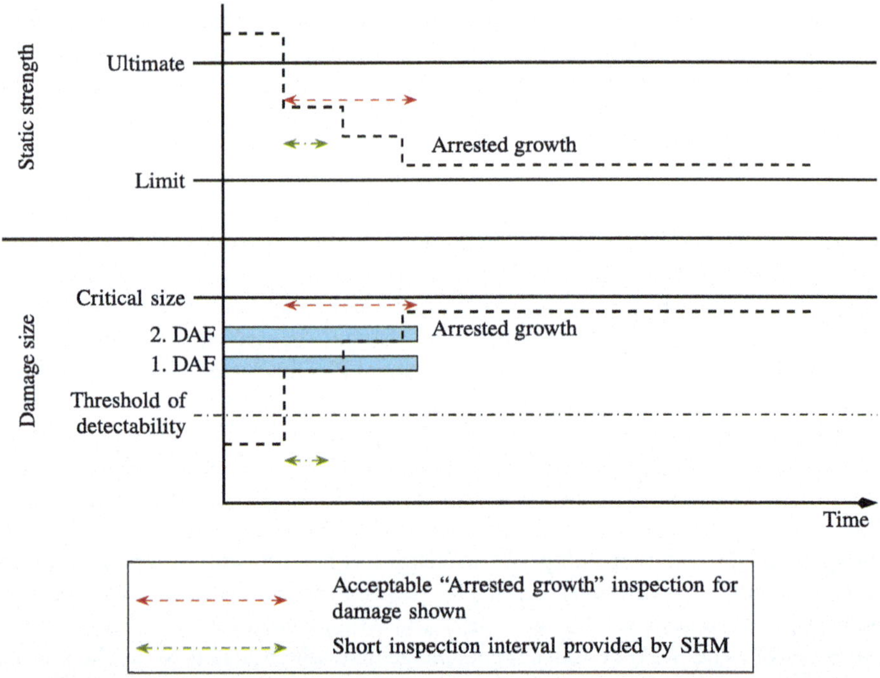

Fig. 4.55 Damage tolerance approach for structural bonds using the two-stage MDAF, depicting the arrested growth method. Adapted from Fig. 3.3

ensures a short inspection interval. Hence, the MDAF is a promising approach to enable the certification of structural bonds by a damage tolerance approach rather than safe-life designs or hybrid joints.

The weight-saving potential of the damage tolerance approach is evidenced by the discussion of Ashcroft [Ash18, pp. 941–942] about how to enable more efficient designs against fatigue failures. The MDAFs seem to meet Ashcroft's condition for controlled crack growth and easily monitoring of fatigue damage for using SHM instead of designing against fatigue failures. As a result, the MDAF is promising for getting rid of the large safety factors, which are necessary when designing against fatigue failures because these are very difficult to predict accurately [Ash18, p. 941]. It has to be stressed, though, that further investigations regarding the performance of the MDAF under environmental degradation of the bondline and more realistic environmental conditions are still necessary. It has to be demonstrated that the aforementioned requirements are still met under these circumstances. The fact that PVDF weakens the bondline rather than strengthens it at temperatures of −55 °C emphasises this necessity.

4.4.2 *Complexity Increase due to the Harmonisation of Competitive Functions*

How does the harmonisation of competitive functions in the multifunctional bondline affect the system complexity?

This section discusses how solving conflicts between competitive functions affects the system complexity.

The first-generation smart inlay MDAF illustrates how the functional integration of the disbond arrest and SHM features causes conflicts during manufacturing. While the disbond arrest function requires a sufficient thickness of the PVDF foil, the micro sensor structures require it to be smooth. These do not seem to be opposing requirements, however as illustrated in [Hei+21, pp. 8–9], the commercial foil with sufficient thickness 100 μm does not meet the smoothness requirements. Moreover, typical processes to produce foils that are suitable for sensor substrates are incapable of providing the required thickness for the DAF. Spin-coating dissolved PVDF, for example, creates smooth foils of thicknesses in the order of 10 μm. Spinning multiple layers on top of each other threatens the integrity of the foil, as previously spun layers might be damaged by the solvents. One solution to this conflict is the temperature treatment of the commercial foil for smoothening it [Hei+22, p. 5], which increases the complexity of the manufacturing process.

The first-generation smart inlay MDAFs also demonstrate the conflicting requirements regarding the stiffness of the PVDF substrate. For the disbond arrest function based on surface toughening, the stiffness has to be much lower compared with the adhesive [Sch24, pp. 154–155]. Regarding the support function of the substrate for the filigree sensor structures, the stiffness of the PVDF is required to be high to protect them against breaking due to high deformations of the PVDF substrate [Hei+22, pp. 7–10]. These requirements are not just conflicting but completely contrary, which demands a more complex design. This must provide the low stiffness in the interface of the MDAF and the bondline, and the high stiffness in the interface between the sensor and the substrate. The approach to providing the disbond arrest function and the sensor support function by encapsulating the sensor gird completely in PVDF is promising in theory because it would only require one additional manufacturing step and does not create a new interface. However, it turns out that the approach to encapsulate the sensors in PVDF causes the new conflict regarding the support function during curing and operation.

The conflict between the support function during curing and operation ultimately leads to an increased system complexity. It is not possible to manufacture and integrate a fully encapsulated PVDF sensor using the standard curing process of the CFRP laminate because the sensor structures break when the PVDF melts. Hence, the additional PEI substrate is introduced to harmonise the disbond arrest and support function. This improves the durability of the sensors, but the additional interfaces in the MDAFs lead to a more complex structure. The extra PEI-sensor-PVDF interface even causes failures in the crack growth test. As a result, this interface needs to be supported or improved. Both measures increase the complexity further. Supporting the PEI-PVDF by the strong CFRP-PVDF interface increases the complexity of the manufacturing process because the sensor structures need to be cut out before encapsulation. Improving the PEI-PVDF interface needs a sophisticated pre-treatment of the foils, which also increases the complexity of the manufacturing process.

In this context, it is an interesting question how the idea of moving the sensors from the adherend-adhesive interface into the adherends to keep the sensors away from the high strains in the DAF affects the complexity (see Sect. 4.2.3). This would enable utilising the strong PEI-epoxy and PVDF-epoxy interfaces only because the functional integration of the disbond arrest and SHM functions would be resolved. On the one hand, this would decrease the complexity because the troublesome interface between the PEI sensor and the PVDF layer is eliminated. On the other hand, this would increase the complexity because the sensors and the DAFs would need to be aligned properly, and the signal transmission and energy supply of the smart inlays would have to be done through the laminate. This would lead to a more complex manufacturing process and a concept for supplying the sensors without disturbing the mechanical properties of the laminate.

In summary, the findings of this section illustrate that even though the disbond arrest and SHM functions are complementing, they can enter a competitive relationship upon functional integration. In the case of the smart inlays, this is because the sub-function of supporting the sensor structures, which is necessary for providing the SHM function, conflicts with the disbond arrest functions, but should be provided by the same function carrier.

4.4.3 Conclusions

The complementing disbond arrest and SHM functions of the MDAF show great potential for a requirement reduction regarding the safe and efficient operation of structural bonds. Within the system boundaries of the CLS crack growth test, the PVDF-based MDAFs prevent crack growth and detect it securely at uncritical sizes. The results illustrate how the DAF only needs to slow down the crack growth enough to enable damage detection by the SHM feature instead of arresting it for the MDAF to fulfil its safety functions. Likewise, the SHM feature only has to identify damage that has grown towards the MDAF instead of providing precise information on the damage initiation and size. Hence, it seems feasible to apply a damage tolerance approach to multifunctional bondlines instead of using conservative safe-life designs or hybrid joints. As a result, the MDAF approach is not only promising in terms of enabling certification by the more efficient damage tolerance approach with lower safety factors but also in terms of a manageable design complexity. However, the performance of the MDAF has to be investigated thoroughly under more realistic environmental conditions and in degraded bondlines to confirm this approach is applicable to primary structural bonds of aircraft.

The investigations further demonstrate how the harmonisation of competing functions increases the system complexity. Different stiffness requirements on the PVDF foil regarding the DAF and the sensor substrate lead to an increased complexity in manufacturing and ultimately the entire design of the smart inlay MDAF by adding a PEI layer for a better yet incomplete harmonisation of these requirements. In this context, it is remarkable that the necessity to introduce an extra function carrier does not only come from the opposing stiffness requirements during the operation of the MDAF. The loss of the support function of the PVDF during manufacturing in a standard curing process prevents a possible harmonisation of the stiffness conflict during operation by encapsulation of the sensor in PVDF. Hence, the transient material properties of the PVDF create a design conflict between the manufacturing and the operation phases. These findings also demonstrate that complementing functions can enter a competitive relationship upon functional integration.

The results lead to the following conclusions regarding Research Hypothesis 3: To start with, the results confirm Research Hypotheses 3.1 and 3.2 within the given system boundaries of the CLS test. The surface toughening effect is capable of providing the disbond arrest functions, which enables a less complex manufacturing process in comparison to the hybrid bondline approach. The combination of the disbond arrest and SHM functions does enable looking for clearly damaged bondline sections between MDAFs without putting the whole structure at risk. Hence, it is feasible to reduce the complexity of the SHM system compared to one that is required to detect the crack initiation and determine the size of damage precisely.

The statements of Research Hypothesis 3 that the complementing characteristics of the disbond arrest and SHM functions enable lower requirements for the individual functions and that this, in turn, enables a complexity reduction, can be confirmed, too. However, the results also clearly indicate that this does not guarantee a low system complexity upon functional integration because the latter might still lead to conflicting requirements of both functions. Hence, despite the complementing nature between both functions, a harmonisation of these conflicting requirements is necessary, which increases the complexity again. As a result, an analysis of whether complementing functions enable lowering the requirements of the individual functions is a promising method to decrease the system complexity, but it does not mean a feature integrating both functions automatically enables a low system complexity.

4.5 Evaluation of the Multifunctional Disbond Arrest Features

Do the proposed MDAFs meet the fundamental requirements?

This section evaluates the proposed MDAFs based on the fundamental requirements from Sect. 3.1.4.

The evaluation is done by checking if it can be said with confidence that a requirement is fulfilled or not. The basis for the evaluation is the experimental results. If

Table 4.5 Evaluation of the functional compliance of the MDAFs in the crack growth test regarding the fundamental requirements from Table 3.1

No.	Requirement	FBG MDAF	Smart inlay MDAFs Gen. 1	Smart inlay MDAFs Gen. 2	Smart inlay MDAFs Gen. 3
1	The residual strength of the bonded structure must substantiate the design load capacities at all times during operation.	✓	✗	✗	?
1.1	Damages within the bondline must be prevented from rapidly growing to a critical size.	✓	✓	?	✓
1.2	Damages within the bondline must be detected before the integrity of the structure is at risk.	✓	✗	✗	✓
1.3	Repeated estimates of the bondline's residual strength during operation must be enabled.	✓	✗	✗	?
1.4	There must be an unambiguous distinction between a damaged and a healthy bondline.	✓	✗	✗	✓
2	The total weight of the bonded structure has to be as light as possible. It must not be heavier than a structure joined conventionally, i.e. with additional mechanical fasteners, at least.	?	?	?	?
3	The multifunctional bondline has to be compatible with standard curing processes of composites.	✓	✓	✓	✓

it is not possible to make a qualified decision with the available information, no answer is given for the according requirement. This is generally the case for the requirement that the multifunctional bondline must not be heavier than hybrid joints because the CLS specimen does not provide a realistic weight estimate for the peripheries of the sensors. In this context, it is only evidenced that the MDAFs function without additional fasteners, which is a pre-condition for fulfilling this requirement. Table 4.5 summarises the evaluation.

The FBG MDAF fulfils all fundamental requirements. In the crack growth tests, it shows a disbond arrest function, and the sensors are capable of providing the data for estimating the residual strength of the bondline repeatedly by identifying failed bondline sections. The only exception is the case when the PVDF and FBG fibre are pulled out of the CFRP. Since only a minority of PVDF DAFs have shown this issue, the FBG MDAF is still considered to fulfil the requirement for a repeatable estimation of the residual strength.

The third-generation smart inlay MDAFs fulfil most requirements as well. However, since its sensor structures fail when they are subjected to the loads near the crack tip, there is a question mark regarding their capability of a repeatable estimation of the residual strength. They fulfil this requirement before the crack reaches the DAF, though, which is why they are capable of detecting the damaged bondline section. In contrast to the FBG MDAF, the sensor failure of the smart inlay is much more probable and caused directly by damage to the sensor structure. It thus depends on the specific application if the sensor has to provide repeated measurements after damage detection. For example, if damage has to be repaired immediately after its detection, there is no need for being able to repeat the measurement because the repair would have been triggered already. However, if only the failure of multiple bondline sections would demand a repair, the sensors must stay functional to inform about further crack growth near the already failed section. Hence, it cannot be determined satisfactorily if the third-generation smart inlay fulfils this requirement or not.

The first-generation smart inlay MDAFs are only capable of fulfilling the requirements regarding preventing the rapid crack growth and the compatibility with standard curing processes. The sensors, however, do not meet any of the damage detection requirements because of their early failures.

Regarding the requirements, the second-generation smart inlay MDAFs show the worst performance. Their sensors do not withstand the cyclic loading, and in the case of the uncut PEI substrate, the disbond arrest function is also in question. The question mark for the requirement to prevent the crack from growing fast is because their MDAFs slow down the crack growth, but do not arrest the crack. Since this might be enough to enable a repair of the structure, it cannot be claimed with confidence whether they hinder the crack growth sufficiently to prevent it from rapidly growing to a critical size.

Based on the evaluation, the FBG MDAFs and, with some limitations, the third-generation smart inlay MDAFs fulfil the basic requirements. Hence, these designs can be considered functionally compliant in the system boundaries of the CLS specimen (overlap joint), and the FBG MDAF would get the highest functional compliance score. At this stage, understanding how these designs achieve functional compliance seems more important than the fact that they are considered functionally compliant or which solution gets the higher score. Nevertheless, the functional compliance evaluation demonstrates the feasibility of implementing MDAFs in structural bonds.

Chapter 5
Towards the Implementation of the Multifunctional Bondline in Composite Aircraft

Is it feasible to implement bondlines with MDAF in composite aircraft?

This chapter discusses what the findings of this thesis tell about the implementation of the proposed two-stage MDAFs in a composite fuselage. It thereby analyses how the change of the system boundaries affects the functional compliance. The basis is the fundamental questions and the initial fuselage concept of Löbel [Löb16, p. 180] for the application of hybrid bondline DAFs in composite fuselages from Sect. 3.1.2.

5.1 Revision of the Fundamental Questions

What answers to the fundamental questions regarding the implementation of multifunctional bondlines follow from the findings of this thesis, and which questions are still open?

This section revises the fundamental questions from Sect. 3.1.2 based on the experimental results from Chap. 4, and discusses the remaining open questions for the implementation of MDAFs in composite aircraft.

J. Steinmetz, *Functional Compliance of Multifunctional Bondlines with Disbond Arrest and Health Monitoring Features for Safe and Efficient Composite Joints*, Mechanics and Adaptronics, https://doi.org/10.1007/978-3-031-72422-0_5

Fundamental Question 1
What properties shall be monitored to assess the bondline status?

The load transfer between the adherends provides the necessary information for identifying damage in the bondline. Section 4.2.2 illustrate that strain monitoring is a promising approach for that purpose. More specifically, monitoring the load transmission in an overlap joint by strain measurements enables the detection of damaged bondline sections. It has to be pointed out though that this approach relies on damage growth from the overlap edges. If damage occurs in the centre of an overlap joint instead, it is likely that there is no significant change in the sensor outputs because the load transfer might be completed before the damage.

Fundamental Question 2
What sensor distribution is necessary for monitoring the chosen properties, reliably?

The required number of sensors for damage detection by the load transfer method depends on the complexity of the damage and the load on the structural bond. The results from Sect. 4.2.2 show that the combination of the SHM system with DAFs allows for a low number of sensors because identifying damaged bondline sections suffices instead of characterising the damage precisely. If there is a similarly well-defined damage propagation as in the CLS specimens, damage detection is even feasible with two sensors integrated into each two-stage MDAF at one side of the bondline.

More complex loads and damage scenarios, however, are expected to require additional information for a reliable SHM system. For example, if the bending loads observed in the CLS specimens caused a strain gradient of similar or even higher magnitude than the damaged bondline under tensile loads, the two sensors would not suffice for distinguishing between both cases. Hence, additional information would be necessary for reliable damage detection.

There are several options to increase the amount of information. One is placing sensors on both sides of the bondline. A second option is combining the information from the surrounding two-stage MDAFs. For example, if there is a damaged bondline section between two two-stage MDAFs, the sensors of both two-stage MDAFs will be affected. Hence, a completely damaged bondline section can be identified with higher confidence by comparing the results of neighbouring two-stage MDAFs. A third option for additional information is utilising other sensors or SHM systems in the composite structure. For example, Lammering et al. [LDG18, p. 4] motivate the application of SHM in CFRP by a fuselage section as well. If the CFRP structures themselves shall be monitored in addition to the bondline, combining the information from both SHM systems seems attractive. The additional information from the other sensors could be used for increasing the robustness of the damage detection in the bondline by providing estimates of the current load on the structure, for example.[1] Finally, including the history of the sensor outputs could provide valuable reference values to distinguish between random and systematic changes in the sensor outputs. This option is particularly interesting to exploit the potential of the DAFs to reduce the required number of sensors. Since the DAF ensures damage does not grow to a critical size quickly, measurements can be repeated and compared to each other. This way, false alarms could be ruled out without risking the structural integrity.

Fundamental Question 3
How to supply the sensors with energy and access their output signal?

The investigation in this thesis does not provide sufficient information for a sophisticated energy supply concept. Nevertheless, there are some conclusions possible from the results regarding the connection of the integrated sensors. These, however, correlate strongly with the manufacturability, and are thus discussed in the according Fundamental Question 5.

Fundamental Question 4
How to combine the structural bonding, disbond arrest, and sensing functions in such a way that they neither disturb each other nor other functions of aircraft?

Section 4.2.4 demonstrates that the proposed MDAFs are a promising method for avoiding serious disturbances of the structural bonding function by the disbond arrest and SHM (sensing) functions. The functional integration of sensors into PVDF DAFs is also feasible without jeopardising the disbond arrest function. However, the integration of the sensors directly into the highly strained PVDF-epoxy adhesive interface causes conflicts in the designs of the smart inlay MDAFs. In particular, the stiffness conflict discussed in Sect. 4.4.2 has to be solved to protect the sensors from failing and to provide the disbond arrest function at the same time.

The experimental results from Sect. 4.2.3 also indicate that there are cases when the failure of the disbond arrest function causes failure to the sensor function as well. This is the case when the CFRP-PVDF interface of the FBG MDAF fails and detaches the FBG fibre, for example. Since the sensor signal is not completely lost, it might be possible to indicate such damage still. If not, this is a serious problem for MDAF designs because the reduction of the requirements for the disbond arrest and SHM functions relies on the fact these safety functions work in parallel. If the failure of one of them leads to immediate failure of the other, this is not the case any more.

For the implementation in a composite fuselage of an aircraft, the experimental results mean that further refinements of the MDAF are necessary. It should be ensured that the damage stays in the PVDF-epoxy adhesive interface by strengthening the

[1] Alternatively, the SHM system in the bondline could also help reduce the number of sensors in the composite without compromising the SHM function.

other interfaces in the MDAF. Moreover, the smart inlays require further adjustments regarding the load capacity of their sensor grids. As discussed at the end of Sect. 4.2.3, a promising method seems to be taking the sensor grids further away from the damage. Since the measurement principle relies on the strain in the adherends, it is a valid option to place the sensors in the laminate instead of the interface between the adherend and the bondline. The surrounding stiff fibres are likely to ensure the sensors are not strained to failure, then. Placing the sensors in the laminate would also make it less likely that an unexpected failure of the DAF could break the sensor as well.

For the application of MDAFs in aircraft, the interaction between its functions with other functions of the aircraft requires further investigations.

Fundamental Question 5
Is the multifunctional bondline manufacturable? (Manufacturing tolerances, sensor integrity, etc.)

The results from Sect. 3.4 demonstrate the feasibility of manufacturing the multifunctional bondline with standard curing processes for bonded composite structures. Moreover, Sect. 4.2.4 illustrates that the smart inlay MDAFs do not affect the bondline thickness and fibre orientation more than co-curing of surface toughening elements alone. Hence, it seems feasible to manufacture MDAFs without causing tolerance issues regarding the bondline thickness. In addition to eliminating the tolerance issues, the curing process of the adhesive is more energy efficient. Since only surface toughening elements are used as DAFs in the CLS specimens, no thermoset welding process is required. Therefore, the curing temperature for the adhesive can be lowered from 180 °C to 130 °C in comparison to the hybrid bondline [Löb+16, p. 232; Löb16, p. 83]. Hence, co-curing the MDAFs enables a rather robust process for the integration of the MDAFs without the need for new manufacturing processes.

Nevertheless, there are some challenges during manufacturing. In particular, the sensors have to be handled with care. This includes the brittle FBG fibres as well as the filigree sensor structures of the smart inlays.

The FBG sensors need to be treated carefully to avoid irreversible damage to the FBG fibre. Therefore, the FBG fibres need to be protected against bending, and a practical method is required for leading the fibres out of the surface or the laminate. The manufacturing process illustrated in Sect. 4.1.3 seems only appropriate for the laboratory scale. Placing the sensors in a PTFE tube outside the overlap, and carefully pulling them out of the strap manually, is time-consuming and error-prone. Moreover, the loose fibres have to be protected throughout the remaining manufacturing steps, including the second curing process for adhesive bonding. With increasing fibre length and number of fibres, this appears very challenging.

The smart inlays promise simpler handling as soon as the sensor structures are completely encapsulated. The manufacturing of the CLS specimens illustrates the feasibility of preparing smart inlays on a PEI foil with PVDF cover before co-curing. The PVDF cover protects the filigree structures when handling the parts. If it is possible to lead the sensors out of the bondline using a path outside the highly strained area around the crack tip, the electrical connections of the smart inlays can be accomplished relatively easily after the structural bonding process as well. In that case, it only has to be ensured that the connector pads are protected. The protection of the connector pad against contamination and damage during handling seems feasible. Most likely, covering them with a protective foil until they are connected is sufficient.

The integration of the smart inlays into the laminate might improve their durability, but it would increase the complexity of manufacturing the structural bond as well. The sensors would require a proper alignment to the DAFs during manufacturing, which complicates manufacturing. Moreover, this option would make it much more complex to electrically contact the sensors. Wire connections would either have to go through the thickness of the laminate or the side of it. Leading the sensor out through the side appears unrealistic because the CFRP cannot be cut where the sensors leave the laminate. Using pins for the electrical connection also seems to complicate the manufacturing process significantly, as the pins have to be led through the laminate. Moreover, the connection of the pins to the smart inlays seems challenging as well. In this context, a wireless sensor system appears to be an attractive alternative to keep the disturbance of the laminate as small as possible.

If the smart inlays are to be integrated into the laminate, it seems appropriate to cover the sensors completely in PEI before the integration. This way, the sensors are electrically isolated from the conductive carbon fibres, and the PEI promises a strong interface to the CFRP. Hence, the disturbance of the mechanical properties of the laminate would be kept at a minimum, or they would even be improved by the PEI [Kyr+21].

Fundamental Question 6
Is there an overall gain in sustainability?

The results illustrate the high potential to improve the efficiency of the structural bonds by implementing MDAFs. The capability of providing a safe joining technology without using mechanical fasteners is evident. Moreover, the possibility of the multifunctional bondline to enable a damage tolerance approach is demonstrated by the results. As discussed in Sect. 4.4.3, this could further improve the structural efficiency because the design could become less conservative. However, additional investigations into environmental degradation are still necessary to confirm these findings at more realistic boundary conditions. Moreover, the evaluation regarding the efficiency has to be expanded to the system boundaries of the entire aircraft and all its life cycle phases. Otherwise, an overall gain in sustainability cannot be assured.

For example, the SHM system comes with additional peripheral equipment for running and evaluating the measurements. This lies outside the system boundary of the overlap joint, but it eliminates some weight-saving potential, nevertheless. In the end, the multifunctional bondline and the peripheral equipment together must not be heavier than a bondline secured with mechanical fasteners.

The results do not enable a qualified assumption of how much weight the peripheral equipment will add upon the implementation of MDAFs in the fuselage, but the electrical connections seem to be a major challenge for lightweight SHM systems. For example, Moix-Bonet et al. [MSW18, p. 471] report that the weight of their piezoelectric sensor network for SHM of a door-surrounding structure is mainly driven by the cable weight. Accordingly, Baker and Scott [BS16, p. 564] name the low weight of FBGs and their potential to incorporate multiple strain sensors in a single fibre as a main advantage of FBGs over electrical resistance strain gauges.

The expected weight penalty due to the peripheral equipment of the SHM system, emphasises the benefit of combining the SHM system with a DAF regarding an increased efficiency. For a low weight and energy consumption of the SHM system, it is of interest to keep the number of sensors to a minimum. Reducing the requirement for the SHM system to identify damaged sections instead of damage sizes by the presence of the DAF enables a minimum of two sensors per two-stage MDAF only. Even though this only holds as long as the conditions are comparable to the CLS specimen, the fundamental principle of the multifunctional bondline to reduce the required number of sensors by controlling the damage growth still applies in practical applications like a composite fuselage of aircraft.

In this context, the requirement for a sustainable aircraft illustrates how the system boundaries affect the evaluation of the alternative MDAF designs. On the one hand, the superior fatigue strength and the lighter connections of the FBGs make them the more promising technology for efficient strain measurements during operation. On the other hand, the smart inlays seem to have advantages over the FBG sensors during manufacturing (see Fundamental Question 5). This means, even though, the FBG sensors have advantages over the smart inlays in the operational phase, the smart inlays might compensate for these in the manufacturing phase and vice versa. Hence, it cannot be concluded from the system boundary of the overlap joint alone, which MDAF design is the optimal one for the implementation in a composite aircraft.

5.2 Extending the Initial Fuselage Concept by Multifunctional Disbond Arrest Features

What do possible implementations of MDAFs in the fuselage of composite aircraft look like?

Based on the results of this thesis, this section illustrates potential applications of the multifunctional bondline in the fuselage of composite aircraft. For that matter, it extends the initial fuselage concept from Sect. 3.1.2 by MDAFs.

Figure 5.1 illustrates how the hybrid bondline DAFs could be replaced by two-stage MDAFs in structural bonds between the fuselage skin and stringers. It shows the fuselage section from the initial concept (Fig. 3.1), but with two-stage MDAFs instead of single-stage hybrid bondline DAFs. In addition, there are no DAFs in the doubler based on the results that suggest MDAFs in the load transmitting adherend are sufficient for a disbond arrest function. The same approach is possible for the joints between the skin and frame. However, for the sake of simplicity, the two-stage MDAF is only shown for the doubler joint between the skin and the stiffeners.

Figure 5.1b shows a smart inlay MDAF. In the depicted doubler joint, the track bridges of the two-stage MDAFs can be led out of the bondline outside the expected crack growth area, which solves one core problem of the smart inlay MDAFs. This is possible because the skin exceeds the doubler. Hence, the MDAFs can be designed wider than the doubler, too, which allows rather simple access to the sensors. In this design, smart inlay MDAFs are also considerably easy to integrate because they can be co-cured to the skin and connected after the bonding process, similar to the manufacturing of the CLS specimens.

Implementing an FBG MDAF instead seems more complex. If the sensor shall point in the crack growth direction, the fibre has to be placed below the doubler. Since the fibre direction cannot be changed as easily as the direction of the track bridges, it is questionable whether the fibres can be led out through the side of the MDAF as well.

Moreover, a challenge regarding the repairability of damaged sections comes into view. The FBG MDAFs along the length of the stiffener could be written on a single fibre to reduce the number of integrated FBG fibres. However, it seems difficult to repair a single bondline section if the surrounding MDAFs use the same FBG fibre because the FBG fibre would have to be cut in-between upon repair. As a result, the other sensors in the healthy sections would lose their connections. Finding a solution that allows for reconnecting the FBG fibre after repair appears complex, as it might require leading the fibre out of the laminate in-between all two-stage MDAFs for installing optical connectors.

In contrast, the smart inlay MDAFs, are replaceable individually because they are connected outside the bondline anyway. This illustrates that the smart inlays MDAFs are an attractive alternative to the FBG sensors, even though they are more sensitive to the high strains in the bondline at present. This is why further development of the prototype smart inlay MDAFs is very attractive for the implementation of MDAF in composite aircraft. A wireless signal transmission and energy supply could ensure their weight penalty is low, for example. If it can be assured that the sensors also function long enough to detect damage that requires a repair, they could harmonise the requirements for a safe and efficient operation, a reasonably complex manufacturing process, and a repairable structural bond very well.

The positioning and exact layout of the two-stage MDAF depends on the specific application. According to the minimal load transfer model, the stages of a single two-stage MDAF have to be slightly further apart than the load transfer length of the bonded structure. The number of two-stage MDAFs and the distance between them has to be determined based on the required strength of a single bondline section. Varying the number of two-stage MDAFs provides some design freedom. For example, it seems possible to include only a single two-stage MDAF if half of the bondline could still substantiate the design loads. In this case, the structure has to be repaired

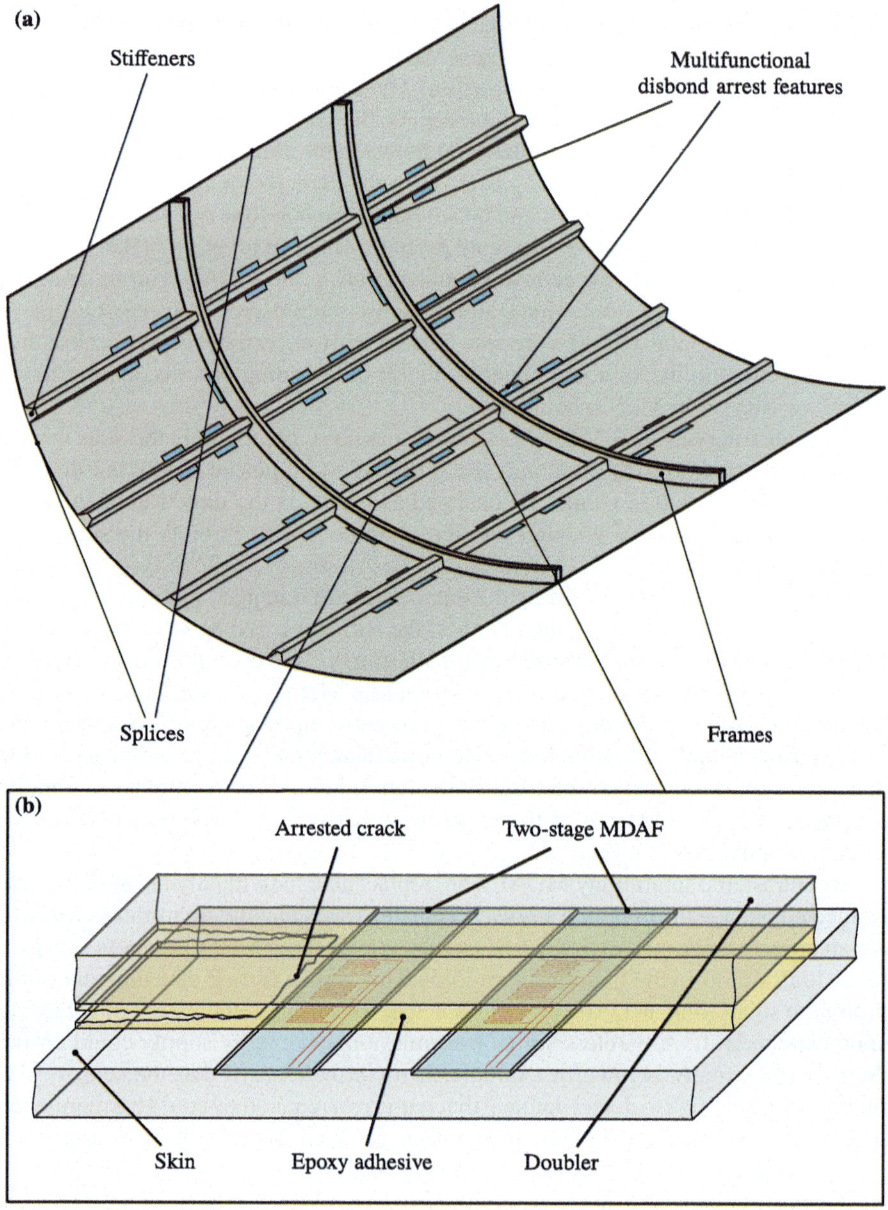

Fig. 5.1 Extension of the fuselage concept with DAFs from Fig. 3.1 by MDAFs **a** Fuselage section with positioning of stiffeners with MDAFs **b** Detailed view of a doubler with smart inlay MDAFs

as soon as the damage is detected. Alternatively, multiple two-stage MDAFs could be integrated. In this case, it has to be determined how many intact bondline sections are required to substantiate the design loads. If the failure of multiple sections is tolerable, the repair of the structure would have to be triggered when the number of failed sections has reached the tolerable amount. The latter scenario could improve the damage tolerance because multiple cracks would have to form before a repair is triggered. However, too many sections would probably be less efficient in terms of weight because of the increased number of sensors and a potentially increased total bondline length. Regarding the minimal distance between the sensors, there always has to be sufficient undisturbed epoxy adhesive in-between multiple two-stage MDAFs to ensure the epoxy adhesive transfers the majority of the load because these are qualified for providing the required stiffness and strength of the bondline. The MDAFs are only meant to improve the damage tolerance, and, if placed at the original overlap edge, to create a better load distribution in the epoxy. They are not meant to, and incapable of, carrying the major share of the load at the required stiffness. Since the aforementioned aspects need to be traded off carefully for selecting an optimal distance between the MDAFs regarding the boundary conditions of the specific application, the number of MDAFs in Fig. 3.1 is arbitrary.

The doubler joints between the skin and stiffeners or frames are of interest because their load profile is similar to the CLS specimens [Löb16, p. 179]. However, these are only classified as secondary joint [BS16, p. 402]. Since one core motivation behind the multifunctional bondline is enabling primary joints based on structural bonding without additional fasteners, a second application scenario Löbel [Löb16, pp. 180–181] proposed for a primary joint is discussed.

The second application scenario is fuselage splices. Kruse et al. [Kru+18, pp. 5–6] developed a Wide Single Lap Shear (WSLS) specimen for testing structural bonds at load conditions representative for high load transfer configurations like longitudinal fuselage splices. The WSLS specimen is taken in the following for discussing the application of MDAFs in a primary joint, which does not share the load profile of a CLS specimen.

Figures 5.2a and b depict a WSLS specimen with an initial disbond in the centre of the joint according to Kruse et al. [Kru+18, p. 4]. The initial disbond represents impact damage or a local weak bond, for example [Kru+18, p. 5]. The numbers one to four indicate the stress peaks in the specimen due to the growth of the disbond, i.e. the size of the disbond increases from state number one to state number four. In contrast to the CLS specimen, the crack growth direction is not in the load direction but perpendicular to it. Moreover, the stress peaks at the crack front exceed the no-growth load level of the specimen, which eventually causes a failure of the structural bond under cyclic loading.

Kruse et al. [Kru+18], thus, study the effect of DAFs made of mechanical fasteners or surface toughening elements on the disbond growth with these WSLS specimens for securing the structural bond. Figure 5.2c depicts the positions of the surface toughening elements. They are placed at the overlap edge in the direction of the disbond growth.

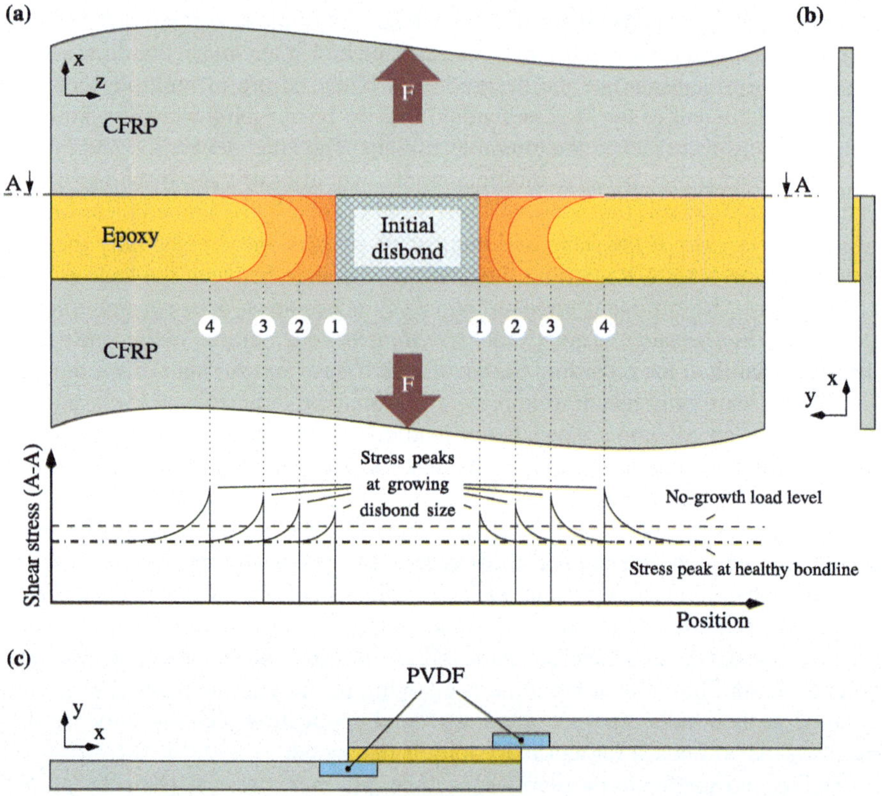

Fig. 5.2 Stress distribution in a WSLS specimen with disbond for testing fuselage splices under realistic damage growth according to [Kru+18, p. 5] **a** Top view, illustrating different disbond length and the corresponding stress peaks in the adhesive along cut A-A **b** Side view **c** Side view of a WSLS specimen with PVDF surface toughening

Both DAFs show an improved resistance against crack growths. The mechanical fasteners arrest the disbond over 3.5×10^7 cycles [Kru+18, p. 8]. The surface toughening DAFs slow down the disbond growth and even keep it below the size of the disbond arrested by the mechanical fasteners for approximately 2.25×10^7 cycles, but afterwards the disbond growths through the entire specimen.

The facts that the healthy bondline is designed for a no-growth scenario, and that the disbond growth is driven by stress peaks at the overlap edges, suggest that the multifunctional bondline could increase the safety and efficiency of this kind of structural bond as well.

The shear stresses illustrated in Fig. 5.2 indicate that there are changing strain gradients due to crack growth in the WSLS, too. Since the location of the stress peak and its amplitude change with the crack propagation, strain gradient measurements in the surface toughening elements seem to be a valid damage indicator. The significantly reduced crack growth in combination with the SHM feature might already suffice for certification. The damage growing perpendicular to the applied load, however, illustrates that the strain distribution is more complex than in the CLS specimen. Hence, further analyses are necessary for identifying the best MDAF positions.

5.3 Conclusions

The implementation of MDAFs for securing structural bonds in the fuselage of composite aircraft is realistic. This is true for the secondary joints between the skin and stiffeners or frames, as well as for the primary joints at the fuselage splices. However, there are still further investigations necessary. For both applications, a more precise analysis of the expected strain distributions is required to determine a suitable MDAF distribution. Most importantly, the performance of the MDAF under other loading conditions and during environmental degradation has to be investigated.

The revision of the fundamental questions and application scenarios demonstrates that a higher functional compliance score in a subsystem does not necessarily mean a better score in the higher system as well. Even though the FBG MDAF shows better results in the experimental investigations of Chap. 4, the smart inlay MDAFs have great potential to become the preferred solution when considering aspects of the higher system like the manufacturability and repairability of fuselage sections with MDAF.

Chapter 6
Extending the Functional Compliance Concept Based on the Multifunctional Bondline

What additional findings can be derived from the investigations of the multifunctional bondline regarding the concept of functional compliance?

This chapter proposes an extension of the functional compliance concept based on the findings from the investigations regarding the multifunctional bondline. It first introduces the concept of temporal and spatial functional compliance and the benefits of considering functional compliance within system boundaries. It then discusses the challenge of evaluating functional compliance in terms of the harmonisation of all functions.

6.1 Global, Local, Temporal, and Spatial Aspects of Functional Compliance

How to account for the dependency of functional compliance on the system boundaries and the transient properties of some function carriers?

The investigations on embedding micro sensors into the bondline of composite structures in Sect. 4.3.1 demonstrate how transient material properties have to be accounted for when harmonising functions. The temporary support function loss of the PVDF sensor substrate is a good example of how changes in the material properties of a function carrier cause the temporary loss of an important function. In this example, the function carrier that is supposed to ensure the sensor integrity even becomes the source of a high deformation of the sensor structure during curing. As a result, the support function has to be provided by a different function carrier to ensure the sensor integrity, and therefore functional compliance upon curing (see Sect. 4.3.1).

J. Steinmetz, *Functional Compliance of Multifunctional Bondlines with Disbond Arrest and Health Monitoring Features for Safe and Efficient Composite Joints*, Mechanics and Adaptronics, https://doi.org/10.1007/978-3-031-72422-0_6

The conflict between the support function of the sensor substrate during curing and the protection of the sensor structure against high strains during operation illustrates another temporal aspect of functional compliance. The findings that encapsulating the PVDF foil sensor would improve its durability during cyclic loading but cause its destruction during curing show how functions that are required in different life cycle phases can be conflicting. This means requirements that do not occur at the same time have to be harmonised to obtain a functionally compliant solution in the end.

Both examples reveal that functions are not only influenced by the spatial distribution of their function carriers but also by changes over time. This leads to the definitions of *temporal* and *spatial* aspects of functional compliance to distinguish between design conflicts originating from the transient behaviour of function carriers and requirements and the design conflicts due to the spatial distribution of the function carriers:

Definition 6.1 *Temporal* functional compliance refers to the harmonisation of all functions in a system, necessary due to the transient behaviour of function carriers or requirements.

Definition 6.2 *Spatial* functional compliance refers to the harmonisation of all functions in a system, necessary due to the disturbances between functions caused by the spatial distribution of their function carriers paired with conflicting requirements.

Distinguishing between spatial and temporal functional compliance does not mean the harmonisation challenge is either a spatial or temporal one. The possible degradation of the epoxy adhesive, for example, illustrates that during operation the spatial and temporal functional compliance regarding the structural bonding, disbond arrest and SHM functions have to be ensured simultaneously because multiple functions require harmonisation at changing material properties.

The distinction between the temporal and spatial aspects of functional compliance emphasises the high complexity accompanied by the objective to obtain functionally compliant systems. The question of how to cope with complex systems brings the great achievements of construction engineering into view. As documented in the VDI 2221, for example, the key idea is breaking down complex systems into manageable subsystems and recombining the solutions of the subsystems to an overall solution [Ver19]. This enables finding solutions for systems so complex that the human brain cannot comprehend them as a whole [Ver19, p. 15]. Hence, this thesis proposes introducing system boundaries to the concept of functional compliance as well. The different meaning of functional compliance regarding the system boundary of an aircraft and the system boundary of an overlap joint illustrates how the system boundaries manage the complexity to meet the design task. This leads to the definition of *local* and *global* functional compliance:

Definition 6.3 *Local* functional compliance refers to functional compliance in a subsystem. *Global* functional compliance refers to functional compliance of a higher-level system created by combining subsystems.

The introduction of system boundaries allows for finding functionally compliant solutions on a subsystem level and aggregating them into an overall solution. Note that this distinction is not limited to two system levels. Any higher-level system can still be a subsystem of another, even higher-level system. Hence, by moving the system boundaries, the complexity of the system can be adjusted as much as necessary to comprehend it. This is true for both the temporal and spatial aspects and leads to the following four combinations:

- *Local, spatial functional compliance* describes the harmonisation of all functions in a subsystem regarding the disturbance between different functions. In terms of the multifunctional bondline, this is the harmonisation of the structural bonding, disbond arrest, and sensing functions for a safe and efficient operation, for example.
- *Local, temporal functional compliance* describes the harmonisation of all functions regarding changes of material properties or requirements within a subsystem. In terms of the multifunctional bondline, this is the harmonisation of the support function with the altering material properties of the surrounding material during curing, for example.
- *Global, spatial functional compliance* describes the harmonisation of all functions regarding disturbances across system boundaries. For the multifunctional bondline, this is the harmonisation of the functions to provide the sensor systems with energy and evaluate the signals that conflict with the requirement of the multifunctional bondline to keep the total weight as low as possible. For example, supplying the sensors through cables could increase the weight significantly and would thus not be considered a functionally compliant solution as soon as the weight penalty exceeds that of hybrid joints, even though the cable system lies outside the subsystem of the multifunctional bondline.
- *Global, temporal functional compliance* describes the harmonisation of all functions across the system boundaries regarding changes in material properties or requirements throughout the product life. Typically, these are design conflicts coming from different requirements in different life cycle phases of a system. For example, the requirement of a strong adhesion during operation can convert into the opposite requirement at the aircraft end of life to be able to separate the components for recycling.

Based on the definition of local and global functional compliance, two cognition hypotheses regarding the design concept of functional compliance can be derived:

Cognition Hypothesis 1: Local and Global Functional Compliance
Local functional compliance is a necessary, but not a sufficient condition for global functional compliance.

Cognition Hypothesis 2: True Global Functional Compliance Optimum
Finding a true global functional compliance optimum is unrealistic because of the simplifications necessary to comprehend complex systems. Getting closer to a true global functional compliance optimum is feasible through iterative optimisations with increased product knowledge.

The aggregation of local solutions does not necessarily lead to a global solution. According to Ziebart [Zie12, pp. 77–81] this is because the reconnection of the subsystems to the overall system might lead to unexpected disturbances. Hence, achieving local functional compliance is no guarantee for global functional compliance. This is in accordance with the statement of Cognition Hypothesis 1. Moreover, Cognition Hypothesis 1 supports the necessity of expanding the definition of functional compliance by a threshold in Definition 3.1 because it allows for selecting multiple functionally compliant designs of subsystems to increase the chance of obtaining global functional compliance.

For the same reason, recombining local solutions might not lead to a global solution, the true global functional compliance optimum is probably never reached. From a construction engineering perspective, the disturbances due to the division and reassembly of a high-level system into comprehensible subsystems will always reduce the solution space [Zie12, pp. 77–81]. Hence, for maximising the solution space, the top-level system, i.e. the system that is not a subsystem itself, would have to be designed as a whole. This, however, is not feasible at present due to the incomprehensible complexity of that task. Consequently, the obtained solutions are always affected by pre-selections that could exclude the true optimum. The Cognition Hypothesis 2, thus, implies that the iterative optimisation process proposed by the VDI 2221 in general and Sinapius [Sin21, p. 2] in terms of functional compliance is crucial for obtaining the best possible solutions. The term functional compliance optimum hence describes the best solution at a time, but never a truly globally optimal solution.

The potential for improvement through iterative optimisation is strongly connected to the aforementioned argumentation of Ziebart [Zie12, pp. 64–66] as well (see Sect. 3.1.1). The need to make high-impact decisions with little knowledge during the early phases of product design decreases the chances of finding optimal solutions from the beginning. Hence, iterations with a more profound understanding of the product most likely lead to improvements. This applies to the concept of functional compliance in particular because this addresses very complex systems with a high degree of functional integration.

The fact that the division into subsystems prevents true design optima sometimes seems to lead to the misunderstanding that this approach would prefer functional separation over functional integration. For example, Pototzky [Pot20, p. 32] claims the VDI 2221 would be unsuited for complex systems like CFRP structures. He argues the VDI 2221 would treat all subsystems completely independently until they are fully designed, which would prevent utilising synergies between function carriers of different subsystems. However, the VDI 2221 explicitly demands an iterative process throughout all phases [Ver19, p. 19], and, it explicitly names an activity for defining modules and their interconnections before their detailed design [Ver19, p. 39]. This activity definitely allows for finding synergies between different solution concepts. The suitability of VDI 2221 for the design of complex systems is further supported simply by the success of the approach in engineering practise. As explained above, the division into subsystems and obtaining solutions for these enables finding solutions for complex systems in the first place.

The VDI 2221 does not prefer functional separation over functional integration or vice versa. This impartiality makes sense because there is no guarantee that strict functional integration or separation usually leads to better solutions. This is supported by the findings in this thesis. For example, the functional integration of the disbond arrest and support functions of the smart inlays by using the PVDF material as substrate and DAF is not functionally compliant because the strain-to-failure of the sensor structures is incompatible with the high deformation of the PVDF required for the disbond arrest function. Since encapsulating the sensor completely in PVDF is also not feasible using standard curing processes, the separation of the support function and the disbond arrest function is necessary. However, this complicates the manufacturing process and leads to a new interface in a critical region. This is a potential threat to the disbond and structural bonding function. From that standpoint, the functional separation between the disbond arrest function and the sensor substrate as done for the FBG sensors might be the preferable solution regarding the requirements for a safe and efficient aircraft. Nonetheless, it is still uncertain which solution would be best regarding the objective of sustainable aircraft, as the smart inlays might allow for a more efficient repair, for example (see Sect. 5.1).

It has to be clarified that even though Pototzky's critique of the VDI 2221 might be misleading, it implies the valid point that a method that helps to find synergies between function carriers is of great interest. His systematic approach to utilising the graph theory for revealing the potential of a material to be used as a function carrier in multiple subsystems [Pot20, Chap. 3] is thus valuable for increasing the number of solutions and obtaining their functional compliance score (see Sect. 6.2). However, the two core challenges of design engineering remain: the tasks are normally too complex to be comprehended as a whole and important decisions have to be made with insufficient product knowledge.

The challenge of making high-impact decisions with little product knowledge points out another interesting design conflict of sustainable aircraft. Sensor integration offers the chance to collect data from practical applications, which helps to close the knowledge gaps for future developments. This causes a conflict regarding the temporal functional compliance. In the long term, precise data on loads, environmental conditions, and everything else that affects the aircraft is of interest to be able to improve the design in the future. Precise data would help to increase efficiency by revealing parts that are designed overly conservatively, for example. Moreover, a more profound understanding of what factors affect the durability of the aircraft would also help to improve the design (see Sect. 3.2.2). In the case of the bondline, this could be a more profound knowledge of how environmental conditions, such as exposure to heat and humidity, affect the structural bond. From that perspective, it would be of interest not to use a simple sensor system as proposed for monitoring the integrity of the multifunctional bondline efficiently but a highly sophisticated one that collects information that tells about the causalities leading to a defect. This stands in conflict with the short-term objective of a sustainable aircraft because additional sensors add weight and increase the energy consumption, for example. This requires a difficult trade-off to achieve optimal sustainability in the short and the long run. A good compromise could even mean that the functional compliance optimum is not the same for all aircraft in a fleet. A valid approach might be to equip some aircraft with more sophisticated sensor systems than others to obtain sufficient data for the future developments while keeping the efficiency of the fleet as high as possible at the same time.

6.2 Evaluating Functionally Compliant Solutions

How to evaluate functionally compliant solutions?

This section deals with the task of evaluating functional compliant solutions for finding the functional compliance optimum. It first explains the challenge regarding evaluating functionally compliant solutions and discusses the established methods, before presenting an approach on how to consider the harmonisation effectively for evaluating functionally compliant solutions.

6.2.1 Challenge of Evaluating Functionally Compliant Solutions

Why is the evaluation of functionally compliant solutions challenging?

A precise evaluation of the functionally compliant solutions is key for finding the functional compliance optimum. However, the demand for harmonising all functions regarding all requirements makes the evaluation of those solutions challenging. The reason for that might be best understood by looking at the main objective of the evaluation process in engineering design.

The objective of evaluation methods is to provide a tool for helping to select a preferred solution from alternative variants [War21, pp. 307–308]. An important reason for using evaluation methods is the demand for a transparent decision. The VDI 2225 explicitly adds the objective to assess the potential for improvement of the evaluated solutions [Ver98, p. 3].

According to Wartzack [War21, pp. 308–309], there are specialised evaluation methods for different phases in design engineering, but the fundamental method is always similar. In the end, this includes retrieving evaluation criteria from the requirements and ranking the solutions based on these. The VDI 2225, for example, proposes a technical evaluation based on the requirements of the system (product) and a scoring system [Ver98, p. 3–7]. A solution principle is only a solution for the system if it fulfils all demands [Ver98, p. 3]. If that is true, minimal requirements, wishes and additional technical properties provide characteristics for a point evaluation [Ver98, p. 3–4]. In the scoring system, each characteristic gets a score between zero (unsatisfying) and four (very good) describing how close the solution is to the ideal solution [Ver98, p. 4]. If possible, this scoring should be done quantitatively, but since many criteria cannot be described by a number, qualitative scores are allowed, too. The total score of the solution is the sum of all points divided by the ideal score. Accordingly, the total score of the solution is a number between zero and one. In addition, the characteristics may be assigned with weights if they shall not have a similar impact on the total score. The characteristics with the least points are the weaknesses of the solution that should be overcome.

In principle, the evaluation method of the VDI 2225 is suited to select the best solution regarding the functional compliance score, too. In fact, the approach proposed in Sect. 3.1.1 to utilise all requirements of the system for evaluating whether a design is a solution and distinguishing between the score of alternative solutions by including minimal requirements and wishes for obtaining the functional compliance score of a system, is completely in agreement with the VDI 2225. Does that mean that as soon as alternative solutions are obtained, the established evaluation methods provide the functional compliance score and optimum, and the task is done? The answer to this is yes and no. Yes, if there are solutions that solve the complex task of harmonising all functions regarding all requirements, their evaluation provides the functional compliance score and optimum. No, because the established evaluation methods do not tell anything explicit about the harmonisation of the requirements, which is a key element of functional compliance. They only tell, if the design is functional and how well it fulfils the requirements.

The value of the functional compliance concept does not seem to be providing the only way to create multifunctional systems with a high amount of functional integration. The value of the functional compliance concept is that it strives to reveal the complementing and competitive aspects of all functions and to foster a systematic understanding of how the requirements for all functions affect each other. The functional compliance concept explicitly asks for the harmonisation of the requirements and an optimisation of it [Sin21, p. 2], which implies an evaluation of how successful the harmonisation is. By doing so, it increases the chances of obtaining and optimising solutions for complex systems with many integrated functions.

The benefit of evaluating the harmonisation of all functions is most obvious when looking at the objective of the evaluation methods to find the potential for improvement of the solutions. Including an evaluation of the harmonisation means gaining knowledge about how adjusting the solution regarding a single function potentially affects other functions. This leads to the systematic understanding of how an improvement in one function of a solution affects the other and a profound understanding of how these coupling effects support and limit the simultaneous improvement of all functions.

The established methods like the VDI 2221 ask for documenting all activities to save the gained knowledge [Ver19, p. 18], but they do not explicitly ask for evaluating how inter-function coupling affects the solutions. Hence, there is always the risk of impairing another function by others or missing out on synergies that could further improve the system. This does not mean the established evaluation methods cannot lead to overall improvements, and that without the functional compliance concept, competitive and complementing functions cannot be identified. But it means that without asking directly for the effect of coupled functions, the chance is higher of missing both positive and negative effects. Even if they are not missed, it means they might not be documented for future developments if they are included by chance rather than cognition. By explicitly looking for coupling effects, the functional compliance concept reveals clear chances and limitations of coupled functions. This raises the question of how to evaluate the harmonisation of all functions.

Analysing how well a solution eliminates the conflicts between all functions appears to be a straightforward approach. However, it can be demonstrated why this approach is ineffective. The criterion implies that a perfect harmonisation removes all conflicts between all functions. This yields two scenarios in which the criterion would be fulfilled best. The first scenario is that the coupled functions are completely complementing anyway. Hence, they would be fully harmonised without any extra effort. The second scenario is that competitive functions are decoupled by the design. This means solutions that functionally separate the conflicting functions would always be fully harmonised, too, because removing the coupling means removing all conflicts between them. Both scenarios are trivial from the perspective of functional compliance because neither requires any harmonisation. Even worse, this criterion would always prefer functional separation as soon as disturbances due to coupled functions occur. This would rule out every solution with a functional integration that causes conflicts, regardless of what the benefits of the functional integration are.

However, the benefits make functional integration valuable. Roth [Rot00, pp. 234–236] discusses the effects of functional integration and separation on the technical and economic value of a system. He illustrates that functional integration tends to decrease the technical and to increase the economic value, while functional separation does the opposite [Rot00, p. 236]. The economic value increases with functional integration because the same number of functions is provided with fewer parts. The technical value decreases with functional integration because of the disturbances between the integrated functions. Roth [Rot00, p. 236] concludes that excessive functional integration and functional separation both reduce the technical and economic value in the end. For evaluating the harmonisation of all functions, this means the technical value alone is not sufficient. Other benefits from functional integration, like the improved economic value, have to be considered as well.

Cognition Hypothesis 3: Evaluating the Harmonisation
An effective harmonisation evaluates the positive and negative effects of functional integration regarding the main objectives, rather than focusing on the elimination of disturbances.

The multifunctional bondline approach illustrates the idea of evaluating the positive and negative effects of functional integration stated in Cognition Hypothesis 3. Regarding the multifunctional bondline, the core objective is increased sustainability. The claim is that by the integration of sensors and DAFs, the sustainability of the bonded structure is increased by fulfilling the safety requirement with a lighter solution than additional mechanical fasteners, which in turn leads to an increased structural efficiency. Moreover, the presence of the SHM system could enable designs with reduced safety factors that allow for weight reduction at the same design loads. However, the potential gain in sustainability is threatened by the conflicts that occur due to the coupling of these functions.

The severe stiffness conflict regarding the support function of the sensor substrate and the disbond arrest function of the DAF in the first-generation smart inlay MDAF results in a dysfunctional sensor, for example (see Sect. 4.4.2). Hence, the high coupling between the two functions leads to a non-functionally compliant system. At this point, the gain in sustainability is irrelevant because the demand for safety is not met.

Reducing the coupling between the sensor and the DAF leads to a functionally compliant solution, but also new disturbances that affect the objective of more sustainable aircraft. Separating the functions of the sensor substrate and the DAF in the second-generation smart inlay MDAFs via the additional PEI phase threatens the disbond arrest function (see Sect. 4.4.2). Hence, a second-generation smart inlay MDAF whose PVDF cover is not in contact with the CFRP laminate and whose track bridge is outside the crack growth area would be only partially functioning. It would momentarily arrest the crack and sense the event of the crack reaching the MDAF. However, the MDAF would be breached eventually, and the sensor would fail when the crack grows over it.

It depends on the safety authorities if this design satisfies the safety demand at all. A possible scenario is that some safety margin would be demanded, which could increase the size of the bondline sections. In that scenario, the sustainability would be affected indirectly by the MDAF because longer bondline sections would require longer adherends, which increases the weight. This means it has to be evaluated whether the indirect loss in sustainability due to the conflict between the disbond arrest and SHM functions is worse than the gain due to removing the mechanical fasteners to select the best solution.

Clearly, the solution of improving the adhesion in the PEI-PVDF interface or supporting it by a CFRP-PVDF interface is more attractive. In that case, the question is how much the manufacturing process is affected to achieve that and how big the impact of this is on the overall sustainability of the aircraft. This also demonstrates that some objectives have to be evaluated across system boundaries, such as the life cycle phases.

The examples illustrate that disturbances can be allowed as long as they do not prevent a requirement from being fulfilled. There are cases when a deliberate disturbance is even necessary for enabling a solution or an overall improvement of the main objectives. This is why an effective harmonisation must evaluate the positive and negative effects regarding the main objectives. This also implies evaluating whether a further improvement of individual functions is accompanied by a gain regarding the main objectives.

This idea is inherent to the multifunctional bondline approach as well. Due to the presence of the DAF, the requirements of the sensor system regarding the objective for a safe bondline are significantly reduced from identifying damage precisely to identifying damaged bondline sections (see Sect. 4.4.1). Hence, the sensor system must be enabled to reliably detect the damaged bondline sections only. Further empowerments, that enable a very precise damage characterisation, seem to add little to the safety of the structure. It is thus more likely that further improving the sensor function harms the objectives more than it serves them. This could either be directly as more sensors could mean more weight or indirectly because the empowerment of the sensor function could impair other functions like the disbond arrest function that affect the objectives more at that point.

Including main objectives changes the question from "Which solution harmonises all functions the best in terms of reducing the disturbances the most?" to "Which solution harmonises the complementing and conflicting functions the best regarding the main objectives?". The advantage of this change of perspective is that it offers the comparison of solutions with different degrees of functional separation and functional integration. Assuming the functional integration leads to an improvement regarding the main objectives, the functional integration approach will be more valuable than the functional separation, as long as the disturbances between the functions do not overcompensate the improvement.

The challenge is thus finding a method that describes the coupling between different functions and how changes in one function affect the main objectives directly and indirectly across all life cycle phases. Such a method would enable a targeted harmonisation of all functions. Moreover, the potential for improvement of each function could be estimated considering its impact on the main objectives, instead of simply assuming that functions with the lowest scores always have the biggest potential for improvement. Such an evaluation method could also indicate when functional separation is better than functional integration.

6.2.2 Evaluating Functional Compliance by Including Coupling Effects via Functional Relations

How could coupling effects and objectives be included effectively in an evaluation process?

The first challenge of this approach is finding a method that considers the coupling of different requirements with each other and the main objectives to describe the effect of the harmonisation on the main objectives. In that context, Behr [Beh23, pp. 2–12] proposes coupling criteria for assessing the competing relations between requirements and coupling matrices for obtaining the functional compliance score. For building the coupling matrix, he utilises a hierarchical structure with the main objective at the top, requirements of these objectives in the middle and criteria for the requirements at the bottom [Beh23, p. 5]. The criteria can be assigned to multiple requirements, which creates the coupling. Pairwise comparisons of the requirements provide the coupling matrix of the requirements for the objective [Beh23, p. 7–8]. The normalised eigenvector of the coupling matrix collects the coupling factors of requirements regarding the objective in a vector. Following the same approach yields the coupling matrices and coupling vectors of every criterion regarding every requirement. Collecting all coupling vectors of the criteria in a matrix, and multiplying the matrix by the coupling vector of the requirements from the right side yields the coupling vector between the criteria and the objective [Beh23, p. 9]. The coupling factors in the vector describe the estimated influence of each criterion on the objective.

The novelty of Behr's method is that it explicitly evaluates how a single criterion influences multiple requirements. This leads to a more detailed determination of how the bottom-level criteria affect the objective, but the method still follows the same principles as the VDI 2225 and other established evaluation methods in the end. Accordingly, Behr [Beh23, pp. 10–11] points out that an evaluation regarding the VDI 2225 would have yielded qualitatively similar coupling values. The reason the results are similar seems to be that Behr acknowledges that a change of a single criterion improves and impairs the fulfilment of different requirements simultaneously [Beh23, p. 6], but the evaluation does not use that important information effectively. The pairwise comparison only tells how strongly the criterion affects different requirements, but it does not tell if increasing the value of a criterion improves or impairs the fulfilment of each requirement it affects. Adding this information appears to be key to an effective evaluation of functionally compliant solutions in terms of the achieved harmonisation.

The need for an expansion of Behr's coupling model is best illustrated by an example. Since there is only a qualitative but no quantitative relation between the surface toughening effect and the properties like stiffness-, strain-, and strength-ratios, a result from Kyriazis [Kyr24] is taken for illustrating the idea of including information about the functional relation between properties and requirements.

Kyriazis [Kyr24] investigates cure monitoring using interdigital foil sensors on a PEI substrate embedded into a CFRP laminate. As illustrated in Fig. 6.1, he obtains a conflict between the shear strength of the laminate and the sensor sensitivity regarding the metallisation degree [Kyr24, p. 166–168]. The metallisation degree is the share of the foil sensor area, which is covered by electrode material. At a metallisation degree between 0 and 28 % there is no influence on the ILSS. With increasing metallisation degree, the ILSS decreases linearly for the tested system (Fig. 6.1b). In contrast, the sensitivity of the interdigital sensor is minimal at 0 and 100 % metallisation degree and maximal at 50 %. There is a nonlinear relation in-between the minima at the side and the maximum in the centre (Fig. 6.1a). Kyriazis [Kyr24, p. 166–168] concludes that there is a Pareto optimum between 28 % and 50 % metallisation degree (red line in Fig. 6.1c).

However, there are additional methods to increase the sensor sensitivity [Kyr24, p. 168]. These are increasing the sensor area or reducing the size of the sensor structure while keeping the metallisation degree constant, for example. Hence, Kyriazis [Kyr24, p. 168] suggests optimising the degree of metallisation for the ILSS and increasing the sensitivity by the other methods.

According to the coupling model of Behr [Beh23, p. 5], the conflict can be described by introducing objectives, requirements, and properties in a hierarchical structure. The objectives are a high load capacity of the structure and the information about the curing status. The requirements are the shear strength and the sensor sensitivity. Since the shear strength only affects the load capacity and the sensitivity only affects the information about the curing status, the requirements are connected to the according objectives. The properties are the metallisation degree, the sensor area, and the size of the sensor structures. Since the metallisation degree affects both requirements, it is connected with both of them, whereas the other two properties are only connected with the sensor sensitivity. All connections are shown in Fig. 6.2.

With Behr's method, the coupling can be illustrated and by using the pairwise comparison, the properties, and requirements could be ranked as explained above. The metallisation degree would probably be evaluated as the most relevant among the three properties, but this information says nothing about how to adjust the properties for harmonising the shear strength and the sensor sensitivity. The evaluation cannot provide this information directly because there is no information on the relation between the properties and the requirements. Kyriazis [Kyr24, p. 167], however, has obtained clear functional relations between the metallisation degree and the shear strength, and the metallisation degree and the sensitivity of the sensors, as illustrated in Fig. 6.1. Using this information enables a qualified statement of what properties should be adjusted.

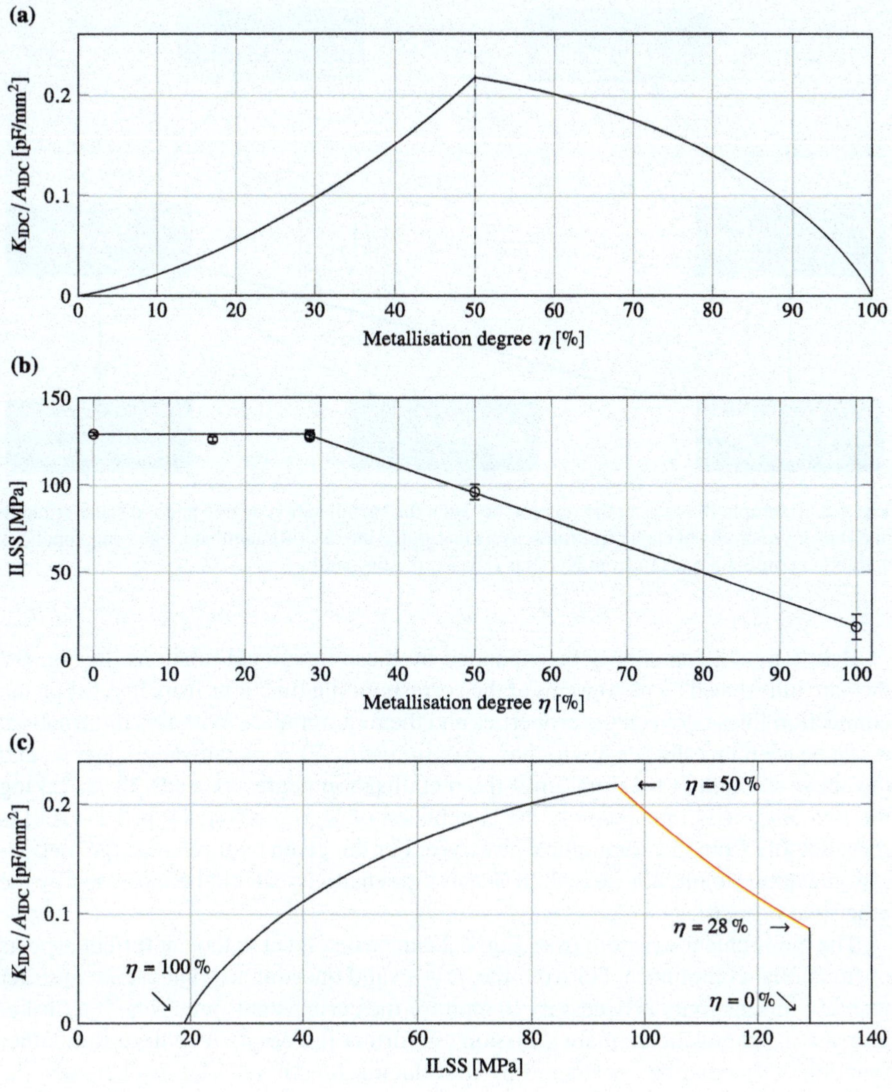

Fig. 6.1 Conflict between the interlaminar shear strength and the areal sensitivity of interdigital foil sensors on PEI substrate integrated into CFRP regarding the metallisation degree η. Reproduced with permission from Kyriazis [Kyr24, p. 167], and the original data provided by the same author. **a** Relation between η and the sensor sensitivity **b** Relation between η and the ILSS **c** Pareto-front of the areal sensitivity and the ILSS

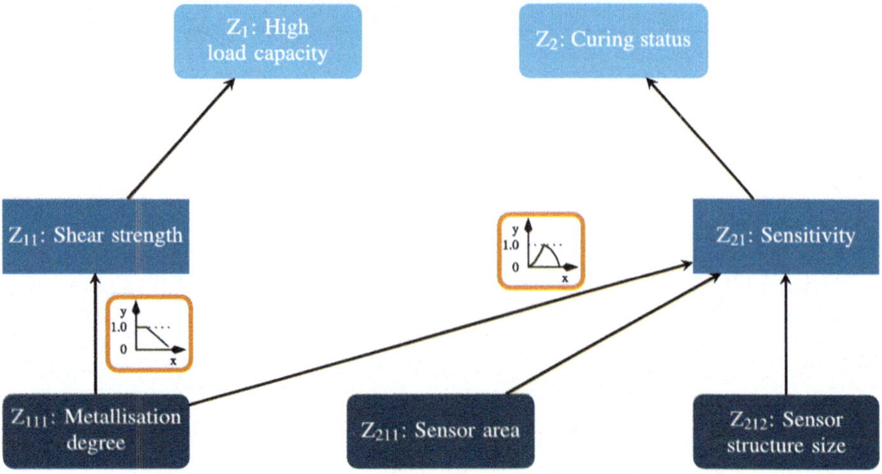

Fig. 6.2 Example illustrating the relation between the metallisation degree and the shear strength and sensor sensitivity of capacitive foil sensors integrated into a CFRP laminate. The value functions and target conflict are taken from Kyriazis [Kyr24, p. 166–168]

Behr's model can simply be expanded by these functional relations. In Fig. 6.2 they are introduced by pictograms of the corresponding functions from Fig. 6.1 on the connection lines between the properties and the requirements. With this information, it can be seen that the sensitivity has its optimum at 50 % metallisation degree, and the shear strength is maximal until the metallisation degree exceeds 28 %. Taking the two properties into account, the conclusion of Kyriazis [Kyr24, p. 166–168] is reproducible from the hierarchical structure: For the given requirements the metallisation degree should not exceed 28 % and the sensitivity should be increased by the other two properties.

The hierarchical structure from Fig. 6.2 can be extended to look at further aspects of functional compliance. For example, it is a valid question to what extent a further increase in sensitivity is necessary to monitor the curing status precisely. If a further increase is not beneficial, there is no point in further increasing it on the cost of other functions. This can be considered by introducing functional relations between the objectives and the requirements as well. Moreover, it is likely that a single objective is not only influenced by one but multiple requirements. If this is the case, it would be of interest to know how much an improvement in a single requirement affects the objective. This could also be done by implementing functional relations between the objectives and the requirements.

Figure 6.3 illustrates the fundamental principle of including coupling and the functional relation between properties, requirements, and objectives for obtaining the functional compliance score. The branches only connect elements from a lower

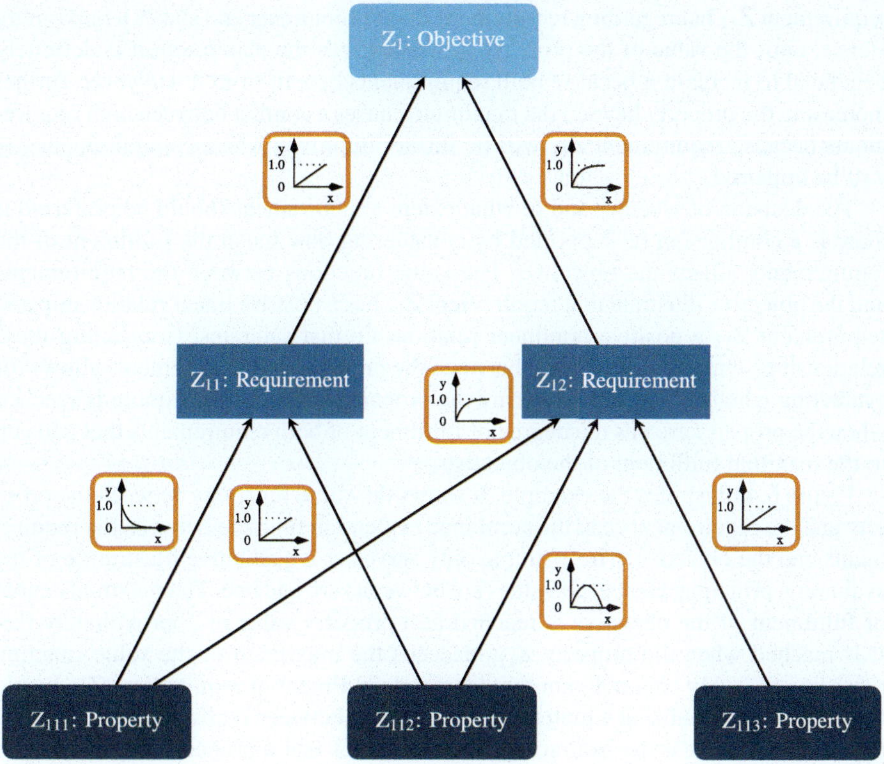

Fig. 6.3 Principal of a hierarchical evaluation of function compliance with value functions on the branches for describing the relations between properties and requirements and the relations between the degree of fulfilment of requirements and the objective

to the next higher hierarchical level. The coupling is included by allowing a connection between a single element and multiple others. For including the functional relations between them, value functions [Zan14, pp. 216–220; War21, pp. 310–331] are included on all connection branches. Preferably, these come from a model or experimental results, but if this is not available, an estimated relation can be considered as well. As demonstrated in the example above, the value functions between the properties and branches describe how an increase in one element affects the element it is connected with. Since the value function does not have to be a positive linear function, the effect of an element on the others can be described more precisely. Even more importantly, an increase in one element can negatively affect another. A conflict regarding changing a property occurs when the value functions of the affected requirements point in different directions.

A brief example considering value functions between all elements illustrates the principle. The question is what the optimal value of property Z_{112} should be. Looking at the value functions of the property Z_{112} first, it is positive and linear regarding requirement Z_{11} but regarding requirement Z_{12} it is nonlinear and shows a maximum. Hence, until the value of the property grows towards the maximum, it is definitely beneficial to increase it because both requirements benefit from it. However, further increasing the property beyond the maximum causes a conflict between both requirements because requirement Z_{11} will be further improved whereas requirement Z_{12} will be impaired.

The decision of whether and to what extent a requirement should be preferred if there is a conflict can be supported by considering how much the fulfilment of the requirements affects the objective. The value functions between the requirements and the objective illustrate that requirement Z_{11} has a positive linear relationship and requirement Z_{12} a positive nonlinear relationship that saturates. Connecting these relationships with the relationship between the property and requirements allows for evaluating whether it is worth altering the property or not. The optimum is reached when the property creates the degree of fulfilment of both requirements that adds up to the maximal fulfilment of the objective.

Figure 6.4 illustrates the example. It shows the value functions between the property and the requirements and the combined evaluation function between the requirements and the objective. The example only serves as a qualitative illustration of the evaluation principle. Hence, all values are between zero and one. The maximal degree of fulfilment of the objective is reached at a property value of approximately 0.8. It is reached when the property Z_{112} exceeds the maximum of the value function regarding Z_{12} only so much that the degree of fulfilment of requirement Z_{12} would be decreased to a value at which the value function between requirement Z_{12} and the objective is still near the saturation. This illustrates that a proper harmonisation of

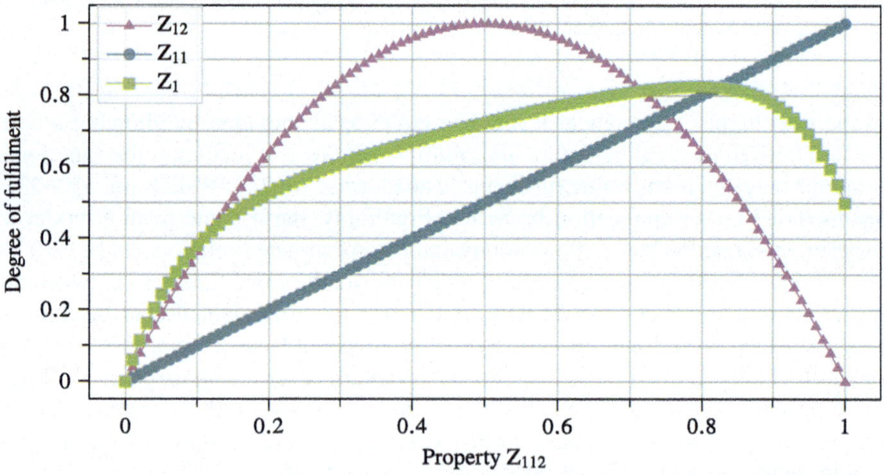

Fig. 6.4 Example how the value functions help determine the harmonisation potential. The curves Z_{11} and Z_{12} show the value functions of the according requirements regarding the property Z_{112}. The curve Z_1 depicts the combined value function of the objective regarding the requirements regarding property Z_{112}

the conflicting requirements would mean impairing requirement Z_{12} slightly more than requirement Z_{11}, as the former has a value of approximately 0.6 and the latter of approximately 0.8 when the objective Z_1 reaches its maximum.

So far, the result for the optimal value of property Z_{112} is obtained without considering the effect of the other properties on the requirements. Taking these into account can change the outcome again because property Z_{113}, for example, only affects requirement Z_{12}. Hence, property Z_{113} could improve the degree of fulfilment of requirement Z_{12} independently of the other properties. Due to the saturating value function between requirement Z_{12} and objective Z_1, property Z_{113} indirectly affects the optimum of property Z_{112}. If property Z_{113} increases the degree of fulfilment of requirement Z_{12}, property Z_{112} would be allowed to cause a bigger impairment of requirement Z_{12} without affecting the objective much. This indicates the complexity of evaluating functionally compliant solutions in terms of the achieved harmonisation.

The example does not illustrate the expected improvement of the objective by introducing the coupling. As a result, Fig. 6.4 gives the impression that the objective cannot reach the theoretical maximum value of one in this example. One might be tempted to conclude that this means again that removing all coupling would offer that possibility, but the simplified example and the shown hierarchical structure are too simplified to account for this comparison. This effect could be accounted for by adding another objective in Fig. 6.3 for representing the positive effect of the introduced coupling. Moreover, complementing functions can reduce some requirements, which could be included in the model by changing the value functions. For example, the value function between the safety objective and the disbond arrest requirement in the multifunctional bondline could be different to the one in a bondline without an SHM system. In the system without SHM it would be a step function because the requirement demands a full arrest throughout the entire service life. In the multifunctional bondline, however, it could be a minimal requirement with a positive relation that saturates. This would indicate that slow growth is acceptable, but a better disbond arrest function improves the safety further until the function also fully arrests the crack.

The hierarchical structure with value functions has the potential to improve the evaluation of functionally compliant solutions by evaluating their harmonisation. One great advantage of the value functions is that experimental data or models provide the most valuable information for expressing the relations. Hence, the model encourages to find these relations and therefore gain a more profound understanding of the interaction between different functions. Moreover, the value functions allow for including temporal functional compliance aspects because the effect of degrading material on the requirements and objectives could be described by the value functions.

The graph method of Pototzky [Pot20, Chap. 3] seems well suited for evaluating functionally compliant solutions regarding their harmonisation, too. The core motivation of Pototzky's graph theory method is the increase in efficiency by the elimination of function carriers (materials) that are redundant because the same function can be fulfilled by other, existing function carriers already [Pot20, p. 25]. Hence, his evaluation method also includes value functions of some sort. Since Pototzky evaluates how much a function carrier contributes to a function, the same approach seems applicable for describing how much a function carrier disturbs a function.

It also seems feasible to apply Pototzky's graph theory method for solving temporal functional compliance problems. Since the graph method collects the material properties and requirements for all function carriers and puts them into relation to find conflicts and synergies [Pot20, pp. 25–30], it seems possible to include the aspects of temporal functional compliance by varying the material properties of the graphs and determining if the functions are still fulfilled when considering all materials. The graph theory method could thus be applied to evaluate a solution considering temporal functional compliance aspects. This implies the relevant transient characteristics of the function carriers are known. As some changes in the characteristics might not be foreseeable but have to be observed first, this loops back to the cognition that engineering design is a highly iterative process.

6.3 Conclusions

The investigations of the multifunctional bondline illustrate that there are spatial and temporal aspects, which affect the functional compliance. The temporal aspects demand a harmonisation of functions due to transient material properties and changing requirements throughout the product life. The spatial aspect demands harmonisation due to the aggregation of multiple functions that disturb each other because of the spatial distribution of their function carriers and conflicting requirements. For a functionally compliant solution, both aspects need to be considered together.

The investigation also demonstrates that introducing system boundaries in the context of functional compliance is important to deal with the system complexity. The proposed concepts of local and global functional compliance help to reduce the system complexity for different design tasks. Since the division in subsystems is accompanied by disturbances upon reconnecting the subsystems, it is wise to look for multiple local solutions in the subsystems to increase the chance of obtaining the most valuable global solution in the end.

Selecting the best possible solution requires sophisticated evaluation methods. The standard methods like the VDI 2225 are suited to determine the best solution among alternative designs. However, they seem incapable of evaluating the potential for improvement efficiently because they do not describe the complementing and conflicting characteristics of system properties on the requirements of different functions. For the evaluation of the harmonisation, it is key to not only consider how well the disturbances of coupled functions are overcome but also how much the gain regarding the main objectives is due to the coupling. If the latter is neglected, there is a high chance that solutions with a high degree of functional separation obtain better scores because they tend to show smaller disturbances, even if solutions with higher disturbances fulfil the objectives better.

Evaluation methods that include value functions for describing the relationship between system properties, requirements, and objectives appear promising for obtaining the functional compliance score considering the achieved harmonisation. They not only show the disturbances due to coupling but also the benefits from it. Moreover, the value functions motivate understanding the relations between functions, which is valuable information for future developments and specific design improvements.

Chapter 7
Conclusions and Future Prospects

What are the most important findings of this thesis, the remaining open questions, and possible next steps?

The closing chapter of this thesis summarises the results, points out the most important findings, and discusses further investigations of the multifunctional bondline itself or related topics. For that matter, this chapter reviews the two main objectives of Sect. 2.2.2, and the fundamental challenge from Sect. 2.1 before coming to the open questions.

7.1 Functional Compliance of the Multifunctional Bondline

> **Central Research Question**
> How to achieve a functionally compliant solution for a multifunctional bondline?

This thesis proposes the introduction of system boundaries for dealing with the complex question of functional compliance regarding the multifunctional bondline. The system boundaries allow for focusing on the fundamental aspects regarding the functional compliance of the main functions of the multifunctional bondline without disregarding the fact that the implementation in an aircraft requires the harmonisation of many more functions. Section 6.1 illustrates that introducing system boundaries is often necessary in design engineering for making a complex design task comprehensible. It thus seems necessary for obtaining functional compliance of complex systems in general, too.

J. Steinmetz, *Functional Compliance of Multifunctional Bondlines with Disbond Arrest and Health Monitoring Features for Safe and Efficient Composite Joints*, Mechanics and Adaptronics, https://doi.org/10.1007/978-3-031-72422-0_7

The experimental investigations in the subsystem of an overlap joint provide valuable information regarding the functionally compliant aggregation of the structural bonding, disbond arrest and SHM functions. These are described regarding the three subsequent research questions in the following.

Research Question 1
How to combine the disbond arrest, health monitoring, and structural bonding functions for a safe operation of bonded structures, while maintaining the mechanical strength and weight-saving potential of an ordinary, undamaged epoxy bondline?

The experimental investigations in Chap. 4 confirm that the implementation of discrete MDAFs in an ordinary epoxy bondline is a valuable approach for combining the structural bonding, disbond arrest, and SHM functions without impairing the strength of the bondline.

The most severe design conflict occurs regarding the stiffness requirements of the DAF and the integrated sensors. For the disbond arrest function, a large deformation of the DAF is required. However, this subjects the integrated sensors to high strains, especially if the crack has grown near the MDAF. Solving this conflict is one major challenge for a functionally compliant design of the smart inlay MDAFs, whereas the FBG fibres withstand the strains.

The investigations illustrate that two steps are important for solving this conflict. These are moving the sensor structures of the smart inlays from inside the PVDF-epoxy adhesive interface to below the interface, and taking the track bridges out of the crack growth area. As a result, the durability of the smart inlays increases sufficiently to detect the event that a crack has reached the MDAF, even though the sensor structures still fails when subjected to the high strains around the crack tip repeatedly. The smart inlay MDAFs and the FBG MDAFs are thus capable of fulfilling the required sensing function for the load-based SHM approach.

The experimental results also illustrate that the surface toughening effect is sufficient for the disbond arrest function (see Sect. 4.2.1). The majority of the PVDF-based (M)DAFs show a disbond arrest effect. Hence, the integrated FBG sensors or micro strain gauges do not impair the disbond arrest function significantly. However, a few (M)DAFs fail in the CFRP-PVDF interface. This not only threatens the disbond arrest function but also the sensor function if the sensor is pulled out of the adherend together with the DAF material. However, these failures are the exception. In contrast, the PEI foil does not show a disbond arrest effect at all because of its greater stiffness.

In terms of the disbond arrest effect, the most important design conflict originates from the smart inlays with a PEI foil as the substrate and a PVDF foil as the cover. There is insufficient adhesion between the PEI substrate and the PVDF layer, which eventually leads to a failure of the interface upon cyclic loading. Supporting the weak interface by bringing the PVDF cover into contact with the CFRP adherend also helps to solve this design conflict. The same applies to the interface between the sensor structure made of gold and the PVDF. The interface itself is weak, but it can be supported sufficiently by the surrounding strong interfaces, like the CFRP-PVDF interface.

The static strength of the structural bond is not impaired significantly by the integrated MDAFs either. This is also attributed to the strong interfaces between PEI or PVDF and CFRP.

Regarding the weight, the bondline itself can be expected to be lighter than a hybrid joint because the materials used for the MDAFs are lighter than mechanical fasteners. However, the system boundary of an overlap joint does not allow for a proper estimate of the weight due to the peripheral equipment of the integrated sensors.

Research Question 2
Is the multifunctional bondline manufacturable with standard curing processes?

The second research question leads to an interesting new aspect of functional compliance. The transient material properties during curing show that not only disturbances between multiple functions require harmonisation but also those due to changing material properties. Since the PVDF melts in standard curing processes of CFRP composites, it loses its function as a support structure for the filigree sensor structures of the smart inlays. Instead, it causes severe deformation of the sensor structures. Hence, these need to be supported by another material with sufficient stiffness to ensure the sensor structures do not fail due to the deformation of the PVDF. This is also why using PVDF as a substrate and cover for the smart inlays is not possible, which leads to another interesting finding regarding the functional compliance of the MDAFs.

From a mechanical perspective, a smart inlay MDAF with the sensor being fully encapsulated in PVDF would probably suffice for detecting the event of a crack having grown to the MDAF. However, it is not manufacturable, which means there is a design conflict that only occurs across two life cycle phases of the structure.

These findings show that functional compliance is not only a matter of harmonising different functions that disturb each other due to their spatial distribution but also of functions that are disturbed by transient material properties or changing requirements. The cognition that there are temporal and spatial aspects of functional compliance (see Sect. 6.1) is thus an important contribution of this thesis.

The challenge of supporting the sensor structures when the adjacent PVDF material melts can be solved by surrounding materials. To protect the sensor structures, ensuring contact with the PEI substrate or the metal tooling is sufficient. Hence, the smart inlay MDAFs are compatible with standard curing processes. The same applies to the FBG MDAFs.

In contrast, there are major challenges regarding the idea to utilise the piezoelectric effect of PVDF despite melting the PVDF. The autoclave pressure does not suffice to maintain the piezoelectric β-phase when the curing temperature exceeds 170 °C (see

Sect. 3.4.3). Moreover, the aforementioned distortion of the micro sensor structures would also affect the electrodes applied to the PVDF. For the piezoelectric sensor, it seems even worse that the electrodes at both sides of the PVDF might get into contact, if the PVDF material in-between melts because this would leave a dysfunctional sensor no matter if a method is found to recover the piezoelectric effect without stretching the PVDF material.

Research Question 3
How to manage the system complexity of the multifunctional bondline?

The final research question leads to interesting new findings regarding the functional compliance as well. The experimental results confirm the assumption from 2.2.2 that the disbond arrest and SHM functions are compatible. This leads to the possibility of decreasing the requirements of both individual functions (see Sect. 4.4.1). This is of great importance for the functional compliance because it allows for utilising the proposed MDAFs in the first place. The entire approach to monitoring the integrity of bondline sections instead of specific damages relies on the requirement reduction. The experimental results support the feasibility of that method.

From these results, another core finding of this thesis follows. The evaluation of functionally compliant solutions has to account for the positive effects of complementing functions when evaluating how successful the harmonisation of a solution is. This might seem obvious, but the coupling of functions is often reduced to the design conflicts. The focus on design conflicts prefers solutions that decouple the functions because decoupling eliminates the disturbances. This, however, disregards the benefits of coupled functions like the potential to fulfil objectives at reduced requirements. This is why it is important to consider the complementing and competing aspects alike when evaluating functional compliance. For that matter, it seems promising to include value functions in the evaluation process that describe the relation between system properties, requirements, and objectives. These allow for assessing if coupling causes a complementing or competitive relationship and if there is a point at which the relationship turns from complementing into competitive or vice versa.

The cognition that controlling the damage growth in the bondline by the DAF simplifies the task for the SHM functions suggests the same principle might be true for other systems, too. It, thus, seems interesting to investigate if the principle of controlling the damage growth for simplifying the SHM system is applicable to other, potentially more complex systems as well. The advantage of the multifunctional bondline is that the damage growth is limited to the bondline, which enables controlling the damage growth by the DAFs. Hence, the damage grows in two dimensions in the worst case.[1] For other application scenarios, the damage control could become a three-dimensional problem, though.

[1] This does not mean damage cannot switch from the bondline to the adherend, but when its growth continues through the laminate, this is a different question than monitoring the bondline integrity.

Main Contributions
This thesis provides the following core contributions regarding the main objective of investigating the feasibility of combining the structural bonding, disbond arrest, and SHM functions in a functionally compliant manner: The results within the system boundary of the overlap joint illustrate that the MDAF approach is promising for obtaining a functionally compliant aggregation of the structural bonding, disbond arrest, and SHM functions. Including the complementing nature of the disbond arrest and SHM functions by adjusting the requirements of both functions accordingly is a key contributor to the functional compliance of the multifunctional bondline.

Regarding the second objective of extending the functional compliance concept, the main contributions are:

- The cognition that applying system boundaries to the functional compliance concept—similar to other engineering design approaches—is necessary to manage complex tasks.
- The understanding that temporal aspects like transient material properties or changing requirements have to be considered for obtaining functionally compliant designs.
- The importance of evaluating the positive and negative effects of coupled functions regarding the core objectives when evaluating how well a design harmonises all functions.

7.2 Efficient Joining Technologies for Composite Aircraft

Fundamental Challenge Regarding Structural Bonding
How to obtain safe structural bonds for joining primary aircraft structures without impairing their lightweight potential?

The investigations of the multifunctional bondline enable some conclusions regarding the fundamental challenge of this thesis about the potential and limitations of structural bonding, too. Figure 7.1 summarises the contribution in that regard.

The results strengthen the assumption from Sect. 2.2.2 that the combination of DAFs and SHM systems is promising for addressing the challenges regarding the implementation of a safe and efficient joining technology based on structural bonding.

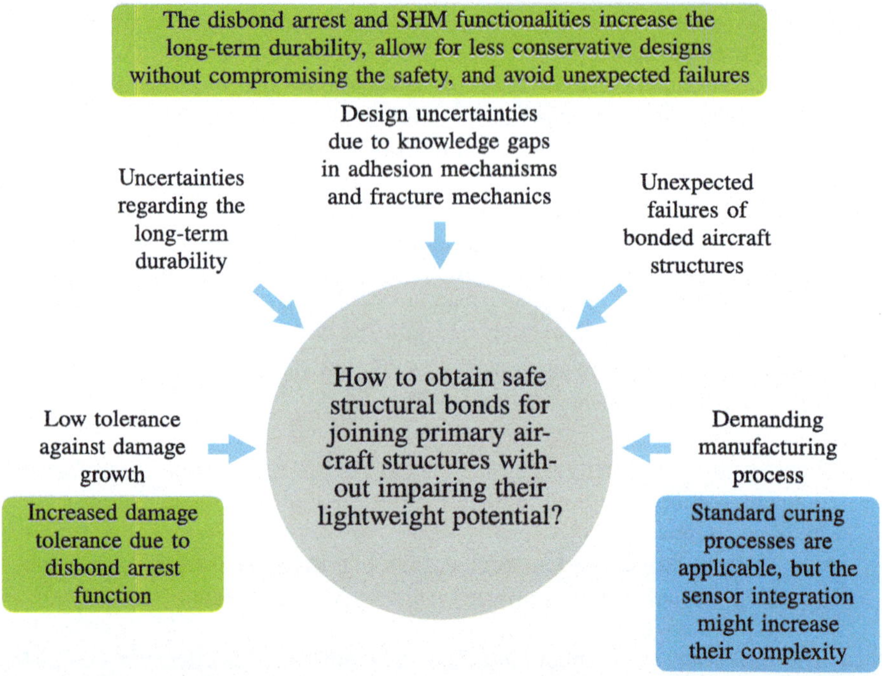

Fig. 7.1 Contribution of this thesis regarding the core challenges for applying structural bonding in primary aircraft structures

The crack growth results from Sect. 4.2.1 illustrate that DAFs based on the surface toughening effect can overcome the low damage tolerance of structural bonds against damage growth.

The results from Sect. 4.2.2 illustrate the capability of the two-stage MDAF to detect damaged bondline sections around the MDAF by strain monitoring and a rather simple load transfer model. In combination with the DAF, the SHM function has great potential to solve the safety concerns originating from uncertainties of the long-term durability of structural bonds and the unexpected failures of bonded structures because it identifies damage to the bondline before it becomes a threat to the structural integrity.

This contributes to the challenge regarding the design uncertainties as well. As illustrated in Sect. 4.4.1, the multifunctional bondline seems to fulfil the requirements for utilising a damage tolerance rather than a safe-life approach for the bondline design because it controls the damage growth and provides short inspection intervals. Hence, the design uncertainties leading to conservative safe-life designs can be overcome indirectly by changing the design philosophy. This might improve the efficiency even more than the potential weight saving by eliminating fasteners because it could reduce the safety factors of the bonded structure and enable maintenance on demand, for example.

Finally, the results from Sect. 4.3 demonstrate that the multifunctional bondline can be manufactured using standard curing processes. This is crucial regarding the challenge of a demanding manufacturing process. Enabling the integration of the MDAF in standard curing processes is thus important in order not to further increase the complexity of the manufacturing process. Although possible, the sensor integration in particular complicates the process. This is especially true for the MDAFs based on FBG sensors because the FBG fibres are damaged quickly, and it appears challenging to protect them in an industrial manufacturing process.

Even though this thesis demonstrates the great potential of the multifunctional bondline, there are still many open questions that require further investigation. The risk that the sensor integration increases the complexity of the already demanding manufacturing process of bonded structures is only one example of how the additional functions could cause disturbances as well.

7.3 Open Questions

What are the most important open questions from this thesis?

Changing the system boundaries from the low level of the structural bond tested under laboratory conditions reveals many open questions regarding the multifunctional bondline approach.

The most fundamental questions regarding the application of the multifunctional bondline follow from including environmental effects. Changing environmental conditions have a major impact on the DAF. Schollerer [Sch24, Chap. 8] illustrates that the PVDF material turns from a surface toughening material into a mechanical disturbance at low temperatures. He thus concludes that PVDF is not suitable for surface toughening at temperatures below −22 °C, which are relevant for aircraft. Accordingly, the fundamental questions of which alternative materials meet the requirements for the application in aircraft and of how environmental effects impact the MDAFs as a whole are important.

The insufficient surface toughening effect at low temperatures directly leads to the next open question of what the exact requirements for obtaining a surface toughening effect are. Until now, the surface toughening effect, regardless if for increasing the static strength or for disbond arrest, has been characterised experimentally mainly. Schollerer [Sch24, Sect. 9.2] provides qualitative design recommendations regarding the stiffness, strain capability, and strength of the surface toughening material, but there is no closed model for a quantitative parameter selection. One reason is that the effect is not completely reproducible in numerical models. Here, Kilian [Kil23, p. 99] shows that including plasticity models in FE simulations improves the prediction of the strength of an SLS joint with surface toughening but still underestimates the strength by at least 20 % compared to experimental results. Being able to predict the surface toughening effect numerically is not only of interest for finding a suitable

DAF material but also for harmonising the stiffness conflict between the sensor and the disbond arrest function.

In the long term, an FE model, which can predict the disbond arrest capability of an MDAF seems very attractive to utilise the approach of Völkerink [Völ22] regarding simulation-driven testing for fatigue as well. This not only requires a model that is capable of predicting the surface toughening effect but also a more profound understanding of how the integrated sensors affect the interfaces. For that matter, DCB and ENF tests with different material combinations of alternative MDAFs are of high interest.

Moving the system boundaries from the overlap joint towards the application in the fuselage of a composite aircraft reveals further important questions regarding the disbond arrest and SHM functions.

Moving beyond the well-defined load case and crack growth direction of the CLS specimen raises the important questions of whether the two-stage MDAF with two measurement points still suffices for arresting cracks and providing unambiguous damage detection. The examples in Chap. 5 illustrate that the load cases might become more challenging and that the damage direction could vary if bondline damage originates from an impact or weak bond, for example. Accordingly, the implementation requires a profound knowledge of the expected load cases, strain distributions, and damage scenarios for selecting a suitable sensor arrangement.

A direct follow-up question is how to ensure the SHM system as a whole does not jeopardise the gain in efficiency by heavy peripheral equipment and cable connections. For that matter, it is an interesting question if self-sustaining MDAFs with wireless data transmission are feasible. In this context, Bornemann [Bor25] presents a self-sufficient, wireless sensor node for SHM using guided waves in fibre metal laminates. Bornemann [Bor25, pp. 17–18] names the vulnerability of wireless approaches to environmental conditions and the detuning of antennas due to conductive materials such as CFRP and metals as two important challenges. The latter renders the integration of the sensor node into fibre metal laminates impossible. Salas et al. [Sal+14, p. 2171], however, illustrate that a power and data transfer to piezo wafer active sensors integrated into CFRP laminates is feasible, depending on the layer geometry and frequency used. Hence, investigating an integrated sensor node with a smart inlay in the future seems very attractive.

For self-sustaining sensor nodes, the complementing nature of the disbond arrest and SHM functions could be of great advantage again. Due to the disbond arrest function, it seems feasible to activate the SHM system only temporarily instead of measuring continuously. This could help enable the self-sustaining MDAFs by reducing their energy consumption. However, this stands in conflict with the idea of measuring repeatedly at the same bondline condition to decrease the chance of false alarms due to random disturbances.

The aforementioned questions mainly relate to the operational phase. For a truly sustainable aircraft, the system boundaries of all life cycle phases have to be accounted for. This raises many more open questions.

From a construction engineering standpoint, the core challenge seems to be accounting for all influences on a product throughout its complete lifetime. The rather new "Integrated Design Engineering" approach is one example of new design methods that aim at accomplishing this and explicitly include sustainability as an integral part of the design process [Vaj22, Chap. 12].

The success of the multifunctional bondline depends on how big its true gain in sustainability throughout the entire product life is and on how well this true gain can be assessed. Understanding the true gain in sustainability is an iterative process because it requires experience and information to understand the full impact of a system. At this point, it should be noted that the system boundaries can lead to an apparent gain in sustainability if important negative effects remain outside the evaluated system boundaries. It is the responsibility of the engineer to use system boundaries appropriately to ensure a true gain in sustainability.

Appendix A
Literature Review Regarding Multifunctional Bondlines

This appendix documents the method for the literature review regarding multifunctional bondlines with disbond arrest and health monitoring functionalities. Table A.1 contains the research protocol, which has led to the following search string:

Scopus Search String SHM and DAF

(TITLE-ABS-KEY (multi*functional OR (function* AND integrat*) OR ("multiple function*") OR ("additional function*") OR ("extra function*") OR concurrent OR simultaneous OR combin*) AND TITLE-ABS-KEY (bond* OR "adhesive joint") AND TITLE-ABS-KEY (dis*bond OR crack OR damage) AND TITLE-ABS-KEY (arrest OR stop* OR retard*) AND TITLE-ABS-KEY ("structural health monitoring" OR shm OR "load monitoring" OR "online damage detection"))

As of the 30th of December 2023, the search leads to 9 results. Three of them are own publications, and only [KRB20] is about an actual application of disbond arrest, SHM and structural bonding. Hence, a search for the authors in Scopus was conducted to see if their research has been expanded since. However, no follow-up research has been published.

J. Steinmetz, *Functional Compliance of Multifunctional Bondlines with Disbond Arrest and Health Monitoring Features for Safe and Efficient Composite Joints*, Mechanics and Adaptronics, https://doi.org/10.1007/978-3-031-72422-0

Table A.1 Research protocol: search terms

Core terms	Multifunctional	Bondline	Disbond	Arrest	Structural health monitoring
Synonyms	Multiple functions, additional functions, extra functions	Adhesive joint, adhesive bond		Stop, retard	SHM, online damage detection
Hypernyms	Functionality	Structural bond, joint	Damage		Damage detection, load monitoring
Related terms	Functional integration, concurrent, simultaneous, combine		Crack		
Foreign terms	Multifunktional, Funktionsintegration	Klebschicht, Klebverbindung	Riss, Schaden		Schadensüberwachung, Strukturüberwachung

Appendix B
Virtual Strain Gauge Measurement

Figure B.1 illustrates the strain field and the VSG signals of the CLS specimen from Fig. 4.37 for a single crack length. In the depicted state, the crack has reached the front of the second DAF. Under loading, this causes the separated part of the lap to bend away from the strap. This in turn causes compressive strains in the VSGs on the lap.

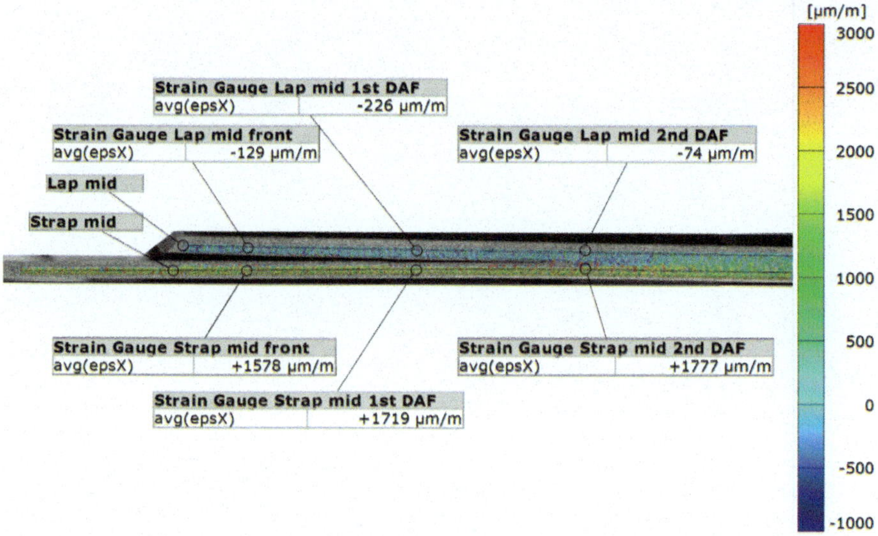

Fig. B.1 Strain distribution and VSG measurements illustrating the bending of a CLS specimen, whose crack has reached the second DAF

J. Steinmetz, *Functional Compliance of Multifunctional Bondlines with Disbond Arrest and Health Monitoring Features for Safe and Efficient Composite Joints*, Mechanics and Adaptronics, https://doi.org/10.1007/978-3-031-72422-0

References

[Ada+16] C. Adams, S. Harput, D. M. J. Cowell, S. Freear, and D. M. Charutz. "Specimen-agnostic guided wave inspection using recursive feedback". In: *2016 IEEE International Ultrasonics Symposium (IUS). 18-21 Sept. 2016*. Ed. by Institute of Electrical and Electronics Engineers. [Piscataway, NJ]: IEEE, 2016, pp. 1–4. ISBN: 978-1-4673-9898-5. DOI: https://doi.org/10.1109/ULTSYM.2016.7728811.

[Agg+21] D. G. Aggelis, M. G. R. Sause, P. Packo, R. Pullin, S. Grigg, T. Kek, and Y.-K. Lai. "Acoustic Emission". In: *Structural Health Monitoring Damage Detection Systems for Aerospace*. Ed. by M. G. R. Sause and E. Jasiūnienė. 1st ed. 2021. Springer Aerospace Technology. Cham: Springer International Publishing and Imprint: Springer, 2021, pp. 175–217. ISBN: 9783030721923. DOI: https://doi.org/10.1007/978-3-030-72192-3_7.

[AM21] I. A. Ashcroft and A. Mubashar. "Fatigue". In: *Adhesive bonding. Science, technology and applications*. Ed. by R. D. Adams. Second edition. Woodhead Publishing series in materials. Duxford, Cambridge, MA, and Kidlington: Woodhead Publishing an imprint of Elsevier, 2021, pp. 317–366. ISBN: 978-0-12-819954-1. DOI:https://doi.org/10.1016/B978-0-12-819954-1.00006-X.

[Ame09] B. Ameduri. "From vinylidene fluoride (VDF) to the applications of VDF-containing polymers and copolymers: recent developments and future trends". eng. In: *Chemical reviews* 109.12 (2009). Journal Article, pp. 6632–6686. ISSN: 0009-2665. DOI: https://doi.org/10.1021/cr800187m.

[AS19] M. Abouhamzeh and J. Sinke. "Effects of fusion bonding on the thermoset composite". In: *Composites Part A: Applied Science and Manufacturing* 118 (2019). PII: S1359835X18305074, pp. 142–149. ISSN: 1359835X. DOI: https://doi.org/10.1016/j.compositesa.2018.12.031.

[Ash18] I. A. Ashcroft. "Fatigue Load Conditions". In: *Handbook of Adhesion Technology*. Ed. by L. F. M. da Silva, A. Öchsner, and R. D. Adams. Second edition. Cham: Springer International Publishing, 2018, pp. 941–973. ISBN: 978-3-319-55410-5. DOI: https://doi.org/10.1007/978-3-319-55411-2_33.

[Bar18] E. J. Barbero. *Introduction to composite materials design*. Third edition. Composite materials. Boca Raton: CRC Press, 2018. 535 pp. ISBN: 9781138196803.

[BE03] M. Benz and W. B. Euler. "Determination of the crystalline phases of poly(vinylidene fluoride) under different preparation conditions using differential scanning calorimetry

J. Steinmetz, *Functional Compliance of Multifunctional Bondlines with Disbond Arrest and Health Monitoring Features for Safe and Efficient Composite Joints*, Mechanics and Adaptronics, https://doi.org/10.1007/978-3-031-72422-0

and infrared spectroscopy". In: *Journal of Applied Polymer Science* 89.4 (2003), pp. 1093–1100. ISSN: 0021-8995. DOI:https://doi.org/10.1002/app.12267.

[Beh23] C. Behr. "Entwurfsmethodik für adaptive Strukturen mit multilateralen Konformitätskriterien". de. Dissertation. Braunschweig: Technische Universität Carolo-Wilhelmina zu Braunschweig, 2023.

[Bor+23] C. S. P. Borges, A. Akhavan-Safar, P. Tsokanas, R. J. C. Carbas, E. A. S. Marques, and L. F. M. da Silva. "From fundamental concepts to recent developments in the adhesive bonding technology: a general view". In: *Discover Mechanical Engineering* 2.1 (2023). PII: 14. DOI: https://doi.org/10.1007/s44245-023-00014-7.

[Bor25] S. Bornemann. *Development of a Self-Sufficient, Wireless Sensor Node Using the Outer FML Layer as an Antenna*. eng. 1st ed. 2025. Mechanics and Adaptronics. Cham: Springer Nature Switzerland and Imprint Springer, 2025. pp. 70. ISBN: 978-3-031-70648-6. https://doi.org/10.1007/978-3-031-70648-6

[BS16] A. A. Baker and M. L. Scott. *Composite materials for aircraft structures*. eng. Third edition. AIAA education series. Reston, VA: American Institute of Aeronautics and Astronautics Inc, 2016. 698 pp. ISBN: 978-1-62410-326-1. DOI: https://doi.org/10.2514/4.103261.

[Cha+13] D. M. Charutz, E. Mor, S. Harput, D. M. J. Cowell, P. R. Smith, and S. Freear. "Guided wave enhancement phased array beamforming scheme using recursive feedback". In: *2013 Joint UFFC, EFTF and PFM Symposium. IEEE-International Ultrasonics Symposium (IUS), Joint IEEE-International Symposium on the Applications of Ferroelectrics (ISAF) and Piezoresponse Force Microscopy and Nanoscale Phenomena in Polar Materials (PFM), Joint IEEE-International Frequency Control Symposium (IFCS) and European Frequency and Time Forum (EFTF) ; 21-25 July 2013, Prague, Czech Republic*. 2013 IEEE International Ultrasonics Symposium (IUS) (Prague, Czech Republic, July 21–25, 2013). International Ultrasonics Symposium et al. Piscataway, NJ: IEEE, 2013, pp. 166–169. ISBN: 978-1-4673-5686-2. DOI: https://doi.org/10.1109/ULTSYM.2013.0043.

[CHN21] R. L. Crane, L. J. Hart-Smith, and J. Newman. "Nondestructive inspection of adhesive bonded joints". In: *Adhesive bonding. Science, technology and applications*. Ed. by R. D. Adams. Second edition. Woodhead Publishing series in materials. Duxford, Cambridge, MA, and Kidlington: Woodhead Publishing an imprint of Elsevier, 2021, pp. 215–256. ISBN: 978-0-12-819954-1. DOI: https://doi.org/10.1016/B978-0-12-819954-1.00008-3.

[DC17] R. Dugnani and F.-K. Chang. "Analytical model of lap-joint adhesive with embedded piezoelectric transducer for weak bond detection". In: *Journal of Intelligent Material Systems and Structures* 28.1 (2017), pp. 124–140. ISSN: 1045-389X. DOI: https://doi.org/10.1177/1045389X16645864.

[DD22] DIN-Normenausschuss Materialprüfung and DIN-Normenausschuss Kunststoffe, eds. *DIN EN ISO 10365:2022-05, Klebstoffe - Bezeichnung der wichtigsten Bruchbilder (ISO _10365:2022); Deutsche Fassung EN _ISO _10365:2022*. DIN EN ISO. Berlin: Beuth Verlag GmbH, 2022. DOI: https://doi.org/10.31030/3305759.

[Dil21] D. A. Dillard. "Applying fracture mechanics to adhesive bonds". In: *Adhesive bonding. Science, technology and applications*. Ed. by R. D. Adams. Second edition. Woodhead Publishing series in materials. Duxford, Cambridge, MA, and Kidlington: Woodhead Publishing an imprint of Elsevier, 2021, pp. 295–316. ISBN: 978-0-12-819954-1. DOI: https://doi.org/10.1016/B978-0-12-819954-1.00014-9.

[dÖA18] L. F. M. da Silva, A. Öchsner, and R. D. Adams. "Introduction to Adhesive Bonding Technology". In: *Handbook of Adhesion Technology*. Ed. by L. F. M. da Silva, A. Öchsner, and R. D. Adams. Second edition. Cham: Springer International Publishing, 2018, pp. 1–7. ISBN: 978-3-319-55410-5. DOI: https://doi.org/10.1007/978-3-319-55411-2_1.

[Ehr+10] B. Ehrhart, B. Valeske, C.-E. Muller, and C. Bockenheimer. "Methods for the Quality Assessment of Adhesive Bonded CFRP Structures - A Resumé". In: *Proceedings of 2nd International Symposium on NDT in Aerospace* 2010 (2010).

[Ehr16] B. Ehrhart. "Quality assessment of bonded primary CFRP structures by means of laser proof testing". Dissertation. Bremen: Universität Bremen, 2016.

[EQF11] D. Elder, B. Qi, and S. Feih. "High strain rate behaviour of bonded composite joints". In: *Composite joints and connections. Principles, modelling and testing / edited by Pedro Camanho and Liyong Tong*. Ed. by P. P. Camanho and L. Tong. Woodhead publishing in materials. Oxford: Woodhead, 2011, pp. 484–507. ISBN: 9781845699901. DOI:https://doi.org/10.1533/9780857094926.2.484.

[Eur10] European Union Aviation Safety Agency. *Annex II to ED Decision 2010/003/R of 19/07/2010*. EASA. 2010.

[Eur20] European Commision, ed. *Communication from the Commission to the European Parliament, the Council, the European Economic and Social Committee and the Committee of the Regions. Sustainable and Smart Mobility Strategy - putting European transport on track for the future*. 2020.

[GdC21] H. R. Gualberto, F. do Carmo Amorim, and H. R. M. Costa. "A review of the relationship between design factors and environmental agents regarding adhesive bonded joints". In: *Journal of the Brazilian Society of Mechanical Sciences and Engineering* 43.8 (2021). PII: 3105. ISSN: 1678-5878. DOI: https://doi.org/10.1007/s40430-021-03105-2.

[Ger+21a] K. Gericke, B. Bender, G. Pahl, W. Beitz, J. Feldhusen, and K.-H. Grote. "Der Produktentwicklungsprozess". In: *Pahl/Beitz Konstruktionslehre. Methoden und Anwendung erfolgreicher Produktentwicklung*. Ed. by G. Pahl, W. Beitz, B. Bender, and K. Gericke. 9. Auflage. Berlin: Springer Vieweg, 2021, pp. 57–93. ISBN: 978-3-662-57302-0. DOI: 10.1007/978-3-662-57303-7_4.

[Ger+21b] K. Gericke, B. Bender, G. Pahl, W. Beitz, J. Feldhusen, and K.-H. Grote. "Grundlagen methodischen Vorgehens in der Produktentwicklung". In: *Pahl/Beitz Konstruktionslehre. Methoden und Anwendung erfolgreicher Produktentwicklung*. Ed. by G. Pahl, W. Beitz, B. Bender, and K. Gericke. 9. Auflage. Berlin: Springer Vieweg, 2021, pp. 27–55. ISBN: 978-3-662-57302-0. DOI: https://doi.org/10.1007/978-3-662-57303-7_3.

[GHO02] B. Giddings, B. Hopwood, and G. O'Brien. "Environment, economy and society: fitting them together into sustainable development". In: *Sustainable Development* 10.4 (2002), pp. 187–196. ISSN: 0968-0802. DOI:https://doi.org/10.1002/sd.199.

[Giu14] V. Giurgiutiu, ed. *Structural Health Monitoring with Piezoelectric Wafer Active Sensors*. eng. 2. ed. Amsterdam: Elsevier and AP Academic Press/Elsevier, 2014. 1012 pp. ISBN: 9780124186910.

[Gru+20] N. Grundmann, H. Brüning, K. Tserpes, T. Strohbach, and B. Mayer. "Influence of Embedding Fiber Optical Sensors in CFRP Film Adhesive Joints on Bond Strength". eng. In: *Sensors (Basel, Switzerland)* 20.6 (2020). Journal Article, p. 1665. ISSN: 1424-8220. DOI: https://doi.org/10.3390/s20061665.

[Guo+09] Y. L. Guo, S. L. Ogin, T. F. Capell, A. M. Thorne, G. T. Reed, Y. S. Wang, and S. C. Tjin. "Effect of Disbond Propagation on the Reflected Spectra of CFBG Sensors Embedded within the Bondline of Composite Bonded Joints". In: *Advanced Materials Research* 79-82 (2009), pp. 2067–2070. ISSN: 1662-8985. DOI: https://doi.org/10.4028/www.scientific.net/AMR.79-82.2067.

[Hab09] G. Habenicht. *Kleben. Grundlagen, Technologien, Anwendungen*. 6th ed. VDI-Buch. Berlin Heidelberg: Springer Berlin Heidelberg, 2009. ISBN: 978-3-540-85266-7.

[Har02] L. J. Hart-Smith. "Adhesive Bonding of Composite Structures—Progress to Date and Some Remaining Challenges". In: *Journal of Composites Technology and Research* 24.3 (2002), pp. 133–153. ISSN: 08846804. DOI: https://doi.org/10.1520/CTR10566J.

[Har18] L. J. Hart-Smith. "Adhesively Bonded Joints in Aircraft Structures". In: *Handbook of Adhesion Technology*. Ed. by L. F. M. da Silva, A. Öchsner, and R. D. Adams.

Second edition. Cham: Springer International Publishing, 2018, pp. 1235–1284. ISBN: 978-3-319-55410-5. DOI: https://doi.org/10.1007/978-3-319-55411-2_44.

[Har21] L. J. Hart-Smith. "Aerospace industry applications of adhesive bonding". In: *Adhesive bonding. Science, technology and applications*. Ed. by R. D. Adams. Second edition. Woodhead Publishing series in materials. Duxford, Cambridge, MA, and Kidlington: Woodhead Publishing an imprint of Elsevier, 2021, pp. 763–800. ISBN: 978-0-12-819954-1. DOI:https://doi.org/10.1016/B978-0-12-819954-1.00001-0.

[Hau18] M. Z. Hauschild. *Life Cycle Assessment*. 1st ed. [Place of publication not identified]: Springer International Publishing, 2018. 1216 pp. ISBN: 978-3-319-56474-6. DOI: https://doi.org/10.1007/978-3-319-56475-3.

[Hau23] A. Hauffe. *expandable Laminate eXplorer*. TU Dresden, Institut für Luft- und Raumfahrttechnik. 2023. URL: https://tu-dresden.de/ing/maschinenwesen/ilr/lft/elamx2/elamx (visited on 01/08/2024).

[Hei+21] C. von der Heide, J. B. S. Steinmetz, M. J. Schollerer, C. Hühne, J. M. Sinapius, and A. Dietzel. "Smart Inlays for Simultaneous Crack Sensing and Arrest in Multifunctional Bondlines of Composites". In: *Sensors (Basel, Switzerland)* 21.11 (2021). PII: s21113852, p. 3852. ISSN: 1424-8220. DOI: https://doi.org/10.3390/s21113852.

[Hei+22] C. von der Heide, J. B. S. Steinmetz, O. Völkerink, P. Makiela, C. Hühne, J. M. Sinapius, and A. Dietzel. "Polyetherimide-Reinforced Smart Inlays for Bondline Surveillance in Composites". In: *Polymers* 14.18 (2022). PII: polym14183816, p. 3816. DOI: https://doi.org/10.3390/polym14183816.

[HH22] L. Heilmann and M. Hoffmann. "On the sensitivity of bondline control specimens with respect to the detection of adhesion defects". In: *The Journal of Adhesion* (2022), pp. 1–28. ISSN: 0021-8464. DOI: https://doi.org/10.1080/00218464.2022.2072733.

[HWW18] L. Heilmann, P. Wierach, and M. Wiedemann. "Proofed Bonding - A Novel Method for Verifying Adhesion in Adhesively Bonded Composite Repairs". In: *18th European Conference on Composite Materials (ECCM-18). 25-28 June 2018*. Ed. by Applied Mechanics Laboratory. Curran Associates, Inc., 2018. ISBN: 978-1-5108-9693-2.

[HYM22] R. Hirakawa, T. Yasuoka, and Y. Mizutani. "Crack-arresting behavior of adhesively bonded joints with a single rectangular convex/concave shape formed on adherend surfaces". In: *The Journal of Adhesion* (2022), pp. 1–26. ISSN: 0021-8464. DOI: https://doi.org/10.1080/00218464.2022.2088282.

[KD16] E. V. Koch and A. Dietzel. "Skin attachable flexible sensor array for respiratory monitoring". In: *Sensors and Actuators A: Physical* 250 (2016), pp. 138–144. ISSN: 09244247. DOI: https://doi.org/10.1016/j.sna.2016.09.020.

[Kil23] R. Kilian. "Erweiterung der numerischen Modellierung von Klebverbindungen um den Surface Toughening Effekt". Institut für Mechanik und Adaptronik. Masterarbeit. Braunschweig: Technische Universität Carolo-Wilhelmina zu Braunschweig, 2023.

[KRB20] M. Kadlec, R. Růžek, and P. Bělský. "Concurrent use of Z-pins for crack arrest and structural health monitoring in adhesive-bonded composite lap joints". In: *Composites Science and Technology* 188 (2020). PII: S0266353819329021, pp. 1–8. ISSN: 02663538. DOI: https://doi.org/10.1016/j.compscitech.2019.107967.

[Kru+18] T. Kruse, T. Koerwien, T. Meer, and M. Geistbeck. "Certification by means of disbond arrest features and results (EU-FP7 project BOPACS)". In: *Use of bonded joints in military applications* STO-MP-AVT-266 (2018).

[Kyr+21] A. Kyriazis, J. Feder, K. Rager, C. von der Heide, A. Dietzel, and M. Sinapius. "Reducing the Weakening Effect in Fibre-Reinforced Polymers Caused by Integrated Film Sensors". In: *Journal of Composites Science* 5.10 (2021). PII: jcs5100256, p. 256. DOI: https://doi.org/10.3390/jcs5100256.

[Kyr24] A. Kyriazis. *Bauteilintegrierte Überwachung variothermer Faserverbundfertigungsprozesse mittels minimalinvasiver Foliensensoren auf löslichem Substrat*. ger. 1st ed. 2024. Mechanics and Adaptronics. Kyriazis, Alexander (VerfasserIn). Berlin, Heidelberg: Springer Berlin Heidelberg and Imprint Springer Vieweg, 2024. 191 pp. ISBN: 978-3-662-68164-0. DOI: https://doi.org/10.1007/978-3-662-68165-7.

[Lan+11] W. Lang, F. Jakobs, E. Tolstosheeva, H. Sturm, A. Ibragimov, A. Kesel, D. Lehmhus, and U. Dicke. "From embedded sensors to sensorial materials—The road to function scale integration". In: *Sensors and Actuators A: Physical* 171.1 (2011). PII: S0924424711002317, pp. 3–11. ISSN: 0924-4247. DOI: https://doi.org/10.1016/j.sna.2011.03.061.

[Lau+18] M. Laurien, B. Demir, H. Büttemeyer, A. S. Herrmann, T. R. Walsh, and L. C. Ciacchi. "Atomistic Modeling of the Formation of a Thermoset/Thermoplastic Interphase during Co-Curing". In: *Macromolecules* 51.11 (2018), pp. 3983–3993. ISSN: 0024-9297. DOI: https://doi.org/10.1021/acs.macromol.8b00736.

[LDG18] R. Lammering, S. Duczek, and U. Gabbert. "Motivation". In: *Lamb-Wave Based Structural Health Monitoring in Polymer Composites*. Ed. by R. Lammering, U. Gabbert, J. M. Sinapius, T. Schuster, and P. Wierach. 1st ed. Research Topics in Aerospace. Cham: Springer International Publishing, 2018, pp. 3–9. ISBN: 978-3-319-49714-3. DOI: https://doi.org/10.1007/978-3-319-49715-0_1.

[Lei+21] W. Leite Cavalcanti, K. Brune, M. Noeske, K. Tserpes, W. M. Ostachowicz, and M. Schlag. *Adhesive Bonding of Aircraft Composite Structures. Non destructive Testing and Quality Assurance Concepts*. Cham: Springer International Publishing, 2021. ISBN: 978-3-319-92809-8. DOI:https://doi.org/10.1007/978-3-319-92810-4.

[Löb+15] T. Löbel, H. Sonnenberg, D. Holzhüter, and C. Hühne. "Fatigue evaluation and enhanced shear strain measurements of bonded composite joints". In: *10th International Conference on Composite Science and Technology*. ICCST/10 (Lisbon). Ed. by A. L. Araúo, J. R. Correia, and C. M. Mota Soares. IST – Instituto Superior Técnico University of Lisbon. 2015, pp. 1–13. ISBN: 978-989-99424-2-4.

[Löb+16] T. Löbel, D. Holzhüter, J. M. Sinapius, and C. Hühne. "A hybrid bondline concept for bonded composite joints". In: *International Journal of Adhesion and Adhesives* (2016), pp. 229–238. ISSN: 01437496. DOI: https://doi.org/10.1016/j.ijadhadh.2016.03.025.

[Löb16] T. Löbel. "The Hybrid Bondline: A Novel Disbond-Stopping Design for Adhesively Bonded Composite Joints". Dissertation. Braunschweig: Technische Universität Carolo-Wilhelmina zu Braunschweig, 2016.

[Lop+11] A. C. Lopes, C. M. Costa, C. J. Tavares, I. C. Neves, and S. Lanceros-Mendez. "Nucleation of the Electroactive Phase and Enhancement of the Optical Transparency in Low Filler Content Poly(vinylidene)/Clay Nanocomposites". In: *The Journal of Physical Chemistry C* 115.37 (2011), pp. 18076–18082. ISSN: 1932-7447. DOI: https://doi.org/10.1021/jp204513w.

[Lov83] A. J. Lovinger. "Ferroelectric polymers". eng. In: *Science (New York, N.Y.)* 220.4602 (1983). Journal Article, pp. 1115–1121. ISSN: 0036-8075. DOI: https://doi.org/10.1126/science.220.4602.1115.

[LP22] W. Li and G. Palardy. "Damage monitoring methods for fiber-reinforced polymer joints: A review". In: *Composite Structures* 299 (2022). PII: S0263822322007899, pp. 1–24. ISSN: 0263-8223. DOI: https://doi.org/10.1016/j.compstruct.2022.116043.

[MLL14] P. Martins, A. C. Lopes, and S. Lanceros-Méndez. "Electroactive phases of poly(vinylidene fluoride). Determination, processing and applications". In: *Progress in Polymer Science* 39.4 (2014), pp. 683–706. ISSN: 00796700. DOI: https://doi.org/10.1016/j.progpolymsci.2013.07.006.

[MOT07] S. Minakuchi, Y. Okabe, and N. Takeda. "Real-time Detection of Debonding between Honeycomb Core and Facesheet using a Small-diameter FBG Sensor Embedded in Adhesive Layer". In: *Journal of Sandwich Structures and Materials* 9.1 (2007), pp. 9–33. ISSN: 1099-6362. DOI: https://doi.org/10.1177/1099636207064457.

[MSW18] M. Moix-Bonet, D. Schmidt, and P. Wierach. "Structural Health Monitoring on the SARISTU Full Scale Door Surround Structure". In: *Lamb-Wave Based Structural Health Monitoring in Polymer Composites*. Ed. by R. Lammering, U. Gabbert, J. M. Sinapius, T. Schuster, and P. Wierach. 1st ed. Research Topics in Aerospace. Cham: Springer International Publishing, 2018, pp. 463–473. ISBN: 978-3-319-49714-3. DOI: https://doi.org/10.1007/978-3-319-49715-0_20.

[MWO15] P. Malinowski, T. Wandowski, and W. Ostachowicz. "The use of electromechanical impedance conductance signatures for detection of weak adhesive bonds of carbon fibre–reinforced polymer". In: *Structural Health Monitoring* 14.4 (2015), pp. 332–344. ISSN: 1475-9217. DOI:https://doi.org/10.1177/1475921715586625.

[MWW22] A. Müller, J. L. Walls, and A. Wittmer. "Sustainable Aviation: An Introduction". en. In: *Sustainable Aviation*. Ed. by J. L. Walls and A. Wittmer. 1st ed. Cham: Springer International Publishing, 2022, pp. 1–22. ISBN: 978-3-030-90894-2. DOI: https://doi.org/10.1007/978-3-030-90895-9_1.

[Nag+22] T. Nagoshi, Y. Harada, S. Nakasumi, N. Yamazaki, K. Hasegawa, K. Takagi, W. Peng, G. Fujii, and M. Ohkubo. "Inherent cohesive failure of epoxy adhesive in carbon-fiber-reinforced plastic composites revealed by micro-tensile testing and finite element analysis". In: *Composites Part B: Engineering* 242 (2022). PII: S1359836822004358, p. 110059. ISSN: 13598368. DOI:https://doi.org/10.1016/j.compositesb.2022.110059.

[Nak+03] K. Nakamura, D. Sawai, Y. Watanabe, D. Taguchi, Y. Takahashi, T. Furukawa, and T. Kanamoto. "Effect of annealing on the structure and properties of poly(vinylidene fluoride) -form films". In: *Journal of Polymer Science Part B: Polymer Physics* 41.14 (2003), pp. 1701–1712. ISSN: 08876266. DOI: https://doi.org/10.1002/polb.10525.

[NI73] K. Nakagawa and Y. Ishida. "Annealing effects in poly(vinylidene fluoride) as revealed by specific volume measurements, differential scanning calorimetry, and electron microscopy". In: *Journal of Polymer Science Part A-2: Polymer Physics* 11.11 (1973), pp. 2153–2171. ISSN: 04492978. DOI: https://doi.org/10.1002/pol.1973.180111107.

[Niu10] M. C. Y. Niu. *Composite airframe structures. Practical design information and data*. eng. 3rd ed. Airframe book series. Hong Kong: Conmilit Press, 2010. 688 pp. ISBN: 9789627128113.

[OJK21] S. Omairey, N. Jayasree, and M. Kazilas. "Defects and uncertainties of adhesively bonded composite joints". en. In: *SN Applied Sciences* 3.9 (2021). PII: 4753, pp. 1–14. ISSN: 2523-3971. DOI: https://doi.org/10.1007/s42452-021-04753-8.

[Ong+18] W. H. Ong, N. Rajic, W. K. Chiu, and C. Rosalie. "Lamb wave–based detection of a controlled disbond in a lap joint". In: *Structural Health Monitoring* 17.3 (2018), pp. 668–683. ISSN: 1475-9217. DOI: https://doi.org/10.1177/1475921717715302.

[Pah+21] G. Pahl, W. Beitz, B. Bender, and K. Gericke, eds. *Pahl/Beitz Konstruktionslehre. Methoden und Anwendung erfolgreicher Produktentwicklung*. 9. Auflage. Berlin: Springer Vieweg, 2021. 1162 pp. ISBN: 978-3-662-57302-0. DOI: https://doi.org/10.1007/978-3-662-57303-7.

[Pal+08] J. Palaniappan, S. L. Ogin, A. M. Thorne, G. T. Reed, A. D. Crocombe, T. F. Capell, S. C. Tjin, and L. Mohanty. "Disbond growth detection in composite–composite single-lap joints using chirped FBG sensors". In: *Composites Science and Technology* 68.12 (2008), pp. 2410–2417. ISSN: 02663538. DOI: https://doi.org/10.1016/j.compscitech.2007.09.020.

[Per11] C. Perrow. *Normal Accidents. Living with High Risk Technologies*. Princeton paperbacks. Princeton: Princeton University Press, 2011. 451 pp. ISBN: 9781400828494. DOI: https://doi.org/10.1515/9781400828494.

[Pet15] R. A. Pethrick. "Design and ageing of adhesives for structural adhesive bonding – A review". In: *Proceedings of the Institution of Mechanical Engineers, Part L: Journal of Materials: Design and Applications* 229.5 (2015), pp. 349–379. ISSN: 1464-4207. DOI: https://doi.org/10.1177/1464420714522981.

[Pin+20] S. Pinheiro Melo, A. Barke, F. Cerdas, C. Thies, M. Mennenga, T. S. Spengler, and C. Herrmann. "Sustainability Assessment and Engineering of Emerging Aircraft Technologies—Challenges, Methods and Tools". In: *Sustainability* 12.14 (2020). PII: su12145663, p. 5663. DOI: https://doi.org/10.3390/su12145663.

[Pot20] A. Pototzky. "Graphenunterstütze Konstruktionsmethode zur Funktionsintegration in Leichtbaustrukturen". de. Dissertation. Braunschweig: Technische Universität Carolo-Wilhelmina zu Braunschweig, 2020.

[Pot79] D. L. Potter. *Primary Adhesively Bonded Structure Technology (PABST). Design Handbook for Adhesive Bonding*. Douglas Aircraft Co Long Beach Ca, 1979.

[Pre+15] A. Preisler, Z. Sadeghi, A. Adomeit, and K.-U. Schröder. "Damage Assessment in Adhesively Bonded Structures by Using SmartSHM". In: *Structural health monitoring 2015. System reliability for verification and implementation ; proceedings of the 10th International Workshop on Structural Health Monitoring ; September 1 - 3, 2015*. Ed. by F.-K. Chang and F. Kopsaftopoulos. Lancaster, Pa.: DEStech Publ, 2015. ISBN: 978-1-60595-275-8. DOI: https://doi.org/10.12783/SHM2015/27.

[Pru+18] R. Prussak, D. Stefaniak, C. Hühne, and J. M. Sinapius. "Evaluation of residual stress development in FRP-metal hybrids using fiber Bragg grating sensors". En;en. In: *Production Engineering* 12.2 (2018). PII: 793, pp. 259–267. ISSN: 1863-7353. DOI: https://doi.org/10.1007/s11740-018-0793-4.

[Ram+20] F. M. Ramírez, M. F. Moura, R. D. Moreira, and F. G. Silva. "A review on the environmental degradation effects on fatigue behaviour of adhesively bonded joints". In: *Fatigue & Fracture of Engineering Materials & Structures* 43.7 (2020), pp. 1307–1326. ISSN: 8756-758X. DOI: https://doi.org/10.1111/ffe.13239.

[RG17] W. Roth and V. Giurgiutiu. "Structural health monitoring of an adhesive disbond through electromechanical impedance spectroscopy". In: *International Journal of Adhesion and Adhesives* 73 (2017), pp. 109–117. ISSN: 01437496. DOI: https://doi.org/10.1016/j.ijadhadh.2016.11.008.

[Rot00] K. Roth. *Konstruieren mit Konstruktionskatalogen*. ger. 3. Aufl., erw. und neu gestaltet. Vol. / Karlheinz Roth ; Bd. 1. Konstruieren mit Konstruktionskatalogen. Berlin, Heidelberg: Springer Berlin Heidelberg, 2000. 440 pp. ISBN: 978-3-642-62099-7. DOI: https://doi.org/10.1007/978-3-642-17466-7.

[Sal+14] M. Salas, O. Focke, A. S. Herrmann, and W. Lang. "Wireless Power Transmission for Structural Health Monitoring of Fiber-Reinforced-Composite Materials". In: *IEEE Sensors Journal* 14.7 (2014), pp. 2171–2176. ISSN: 1530-437X. DOI: https://doi.org/10.1109/JSEN.2014.2305578.

[Sch+15] T. A. Schmid Fuertes, T. Kruse, T. Körwien, and M. Geistbeck. "Bonding of CFRP primary aerospace structures – discussion of the certification boundary conditions and related technology fields addressing the needs for development". In: *Composite Interfaces* 22.8 (2015), pp. 795–808. ISSN: 0927-6440. DOI: https://doi.org/10.1080/09276440.2015.1077048.

[Sch+19] M. J. Schollerer, J. Kosmann, O. Völkerink, D. Holzhüter, and C. Hühne. "Surface toughening – a concept to decrease stress peaks in bonded joints". In: *The Journal of Adhesion* 95.5-7 (2019), pp. 495–514. ISSN: 0021-8464. DOI: https://doi.org/10.1080/00218464.2018.1555041.

[Sch+20] M. J. Schollerer, J. Kosmann, D. Holzhüter, C. Bello-Larroche, and C. Hühne. "Surface toughening – An industrial approach to increase the robustness of pure adhesive joints with film adhesives". In: *Proceedings of the Institution of Mechanical Engineers, Part G: Journal of Aerospace Engineering* 234.13 (2020), pp. 1980–1987. ISSN: 0954-4100. DOI: https://doi.org/10.1177/0954410020950071.

[Sch07] H. Schürmann. *Konstruieren mit Faser-Kunststoff-Verbunden*. ger. 2., bearbeitete und erweiterte Auflage. VDI-Buch. Berlin, Heidelberg: Springer-Verlag Berlin Heidelberg, 2007. ISBN: 978-3-540-72189-5. DOI: https://doi.org/10.1007/978-3-540-72190-1.

[Sch24] M. J. Schollerer. "Steigerung der Robustheit von strukturellen Verklebungen in der Luftfahrt mittels lokaler Oberflächenzähmodifikation". de. Dissertation. Braunschweig: Technische Universität Carolo-Wilhelmina zu Braunschweig, 2024. 198 pp.

[Sin21] J. M. Sinapius. *Adaptronics – Smart Structures and Materials*. Berlin, Heidelberg: Springer Berlin Heidelberg, 2021. ISBN: 978-3-662-61398-6. DOI: https://doi.org/10.1007/978-3-662-61399-3.

[SSW13] D. Schmidt, J. M. Sinapius, and P. Wierach. "Design of mode selective actuators for Lamb wave excitation in composite plates". In: *CEAS Aeronautical Journal* 4.1 (2013), pp. 105–112. ISSN: 1869-5582. DOI: https://doi.org/10.1007/s13272-012-0059-3.

[ST23] I. Sioutis and K. Tserpes. "A Literature Review on Crack Arrest Features for Composite Materials and Composite Joints with a Focus on Aerospace Applications". In: *Aerospace* 10.2 (2023). PII: aerospace10020137, p. 137. DOI: https://doi.org/10.3390/aerospace10020137.

[Ste+21] J. B. S. Steinmetz, T. Löbel, O. Völkerink, C. Hühne, J. M. Sinapius, C. von der Heide, and A. Dietzel. "The Working Principles of a Multifunctional Bondline with Disbond Stopping and Health Monitoring Features for Composite Structures". In: *Journal of Composites Science* 5.2 (2021). PII: jcs5020051, p. 51. DOI: https://doi.org/10.3390/jcs5020051.

[SY03] A. Salimi and A. A. Yousefi. "Analysis Method: FTIR studies of -phase crystal formation in stretched PVDF films". In: *Polymer Testing* 22.6 (2003). PII: S0142941803000035, pp. 699–704. ISSN: 01429418. DOI: https://doi.org/10.1016/S0142-9418(03)00003-5.

[TLd22] A. F. G. Tenreiro, A. M. Lopes, and L. F. M. da Silva. "A review of structural health monitoring of bonded structures using electromechanical impedance spectroscopy". In: *Structural Health Monitoring* 21.2 (2022), pp. 228–249. ISSN: 1475-9217. DOI: https://doi.org/10.1177/1475921721993419.

[Tor13] E. Torenbeek. *Advanced aircraft design. Conceptual design, analysis, and optimization of subsonic civil airplanes*. Aerospace series. Chichester, West Sussex, United Kingdom: Wiley, 2013. ISBN: 9781118568118. DOI: https://doi.org/10.1002/9781118568101.

[TZP18] S. Teixeira de Freitas, D. Zarouchas, and J. A. Poulis. "The use of acoustic emission and composite peel tests to detect weak adhesion in composite structures". In: *The Journal of Adhesion* 94.9 (2018), pp. 743–766. ISSN: 0021-8464. DOI: https://doi.org/10.1080/00218464.2017.1396975.

[US 10] U.S. Department of Transportation Federal Aviation Administration. *Advisory Circular 20-107B - Composite Aircraft Structure*. 2010.

[Vaj22] S. Vajna, ed. *Integrated Design Engineering. Interdisziplinäre und ganzheitliche Produktentwicklung*. ger. 2. Auflage. Berlin and Heidelberg: Springer Vieweg, 2022. ISBN: 978-3-662-60438-0. DOI: https://doi.org/10.1007/978-3-662-60439-7.

[Van+15] P. A. Vanniamparambil, R. Carmi, F. Khan, J. Cuadra, I. Bartoli, and A. Kontsos. "An active–passive acoustics approach for bond-line condition monitoring in aerospace skin stiffener panels". In: *Aerospace Science and Technology* 43 (2015). PII: S1270963815001005, pp. 289–300. ISSN: 1270-9638. DOI: https://doi.org/10.1016/j.ast.2015.03.007.

[Van+21] S. Vanlanduit, M. Sorgente, A. R. Zadeh, A. Güemes, and N. Faisal. "Strain Monitoring". In: *Structural Health Monitoring Damage Detection Systems for Aerospace*. Ed. by M. G. R. Sause and E. Jasiūnienė, 1st ed. 2021. Springer Aerospace Technology. Cham: Springer International Publishing and Imprint: Springer, 2021, pp. 219–241. ISBN: 9783030721923. DOI: https://doi.org/10.1007/978-3-030-72192-3_8.

[Vas+21] N. Vasic, J. B. S. Steinmetz, M. Görke, J. M. Sinapius, C. Hühne, and G. Garnweitner. "Phase Transitions of Polarised PVDF Films in a Standard Curing Process for Composites". eng. In: *Polymers* 13.22 (2021). Journal Article. DOI:https://doi.org/10.3390/polym13223900.

[Vas21] N. Vasic. "Der Einfluss von Wärmebehandlungen unter Einfluss elektrischer Felder auf die kristallinen Phasen von -phasigem PVDF". Institut für Mechanik und Adaptronik. Masterarbeit. Braunschweig: Technische Universität Carolo-Wilhelmina zu Braunschweig, 2021.

[Ver19] Verein Deutscher Ingenieure e.V., ed. *VDI 2221 Blatt 1:2019-11 Entwicklung technischer Produkte und Systeme - Modell der Produktentwicklung*. VDI-Richtlinie. Berlin: Beuth Verlag GmbH, 2019.

[Ver98] Verein Deutscher Ingenieure e.V., ed. *VDI 2225 Blatt 3:1998-11, Konstruktionsmethodik - Technisch-wirtschaftliches Konstruieren - Technisch-wirtschaftliche Bewertung*. VDI-Richtlinie. Berlin: Beuth Verlag GmbH, 1998.

[Völ22] O. Völkerink. "Simulation-driven design of bonded joints in fibre composite aircraft structures using progressive damage analyses". Fakultät für Maschinenbau. en. Dissertation. Braunschweig: Technische Universität Carolo-Wilhelmina zu Braunschweig, 2022. 308 pp.

[War21] S. Wartzack. "Auswahl- und Bewertungsmethoden". In: *Pahl/Beitz Konstruktionslehre. Methoden und Anwendung erfolgreicher Produktentwicklung*. Ed. by G. Pahl, W. Beitz, B. Bender, and K. Gericke. 9. Auflage. Berlin: Springer Vieweg, 2021, pp. 307–334. ISBN: 978-3-662-57302-0. DOI: https://doi.org/10.1007/978-3-662-57303-7_11.

[Wei+20] J. Weiland, D. F. Hesser, W. Xiong, A. Schiebahn, B. Markert, and U. Reisgen. "Structural health monitoring of an adhesively bonded CFRP aircraft fuselage by ultrasonic Lamb Waves". In: *Proceedings of the Institution of Mechanical Engineers, Part G: Journal of Aerospace Engineering* 234.13 (2020), pp. 2000–2010. ISSN: 0954-4100. DOI:https://doi.org/10.1177/0954410020950511.

[Wor87] World Commission on Environment and Development, ed. *Our Common Future*. Report of the World Commission on Environment and Development. United Nations, 1987.

[YAL21] A. Yudhanto, M. Alfano, and G. Lubineau. "Surface preparation strategies in secondary bonded thermoset-based composite materials: A review". In: *Composites Part A: Applied Science and Manufacturing* 147 (2021). PII: S1359835X21001640, pp. 2–23. ISSN: 1359835X. DOI: https://doi.org/10.1016/j.compositesa.2021.106443.

[YRR10] F. Yan, R. L. Royer, and J. L. Rose. "Ultrasonic Guided Wave Imaging Techniques in Structural Health Monitoring". In: *Journal of Intelligent Material Systems and Structures* 21.3 (2010), pp. 377–384. ISSN: 1045-389X. DOI: https://doi.org/10.1177/1045389X09356026.

[YWS17] S. Yashiro, J. Wada, and Y. Sakaida. "A monitoring technique for disbond area in carbon fiber–reinforced polymer bonded joints using embedded fiber Bragg grating sensors: Development and experimental validation". In: *Structural Health Monitoring* 16.2 (2017), pp. 185–201. ISSN: 1475-9217. DOI: https://doi.org/10.1177/1475921716669979.

[Zan14] C. Zangemeister. *Nutzwertanalyse in der Systemtechnik. Eine Methodik zur multidimensionalen Bewertung und Auswahl von Projektalternativen*. Teilw. zugl.: Berlin, Univ., Diss., 1970. ger. 5., erw. Aufl. Norderstedt: Books on Demand, 2014. 414 pp. ISBN: 978-3-923264-00-1.

[Zhu+18] Y. Zhuang, F. Kopsaftopoulos, R. Dugnani, and F.-K. Chang. "Integrity monitoring of adhesively bonded joints via an electromechanical impedance-based approach". In: *Structural Health Monitoring* 17.5 (2018), pp. 1031–1045. ISSN: 1475-9217. DOI: https://doi.org/10.1177/1475921717732331.

[Zie12] J. R. Ziebart. *Ein konstruktionsmethodischer Ansatz zur Funktionsintegration*. Zugl.: Braunschweig, Techn. Univ., Diss., 2012. ger. 1. Aufl. Vol. 83. Bericht / Institut für Konstruktionstechnik, Technische Universität Braunschweig. München: Dr. Hut, 2012. 243 pp. ISBN: 9783843905671.

The manufacturer's authorised representative in the EU is Springer Nature Customer Service Centre GmbH, Europaplatz 3, 69115 Heidelberg, Germany. If you have any concerns regarding our products, please contact ProductSafety@springernature.com

Printed and bound by CPI Group (UK) Ltd, Croydon, CR0 4YY
30/09/2026
02225454-0001